Studies in Systems, Decision and Control

Volume 246

The series "Studies in Systems, Decision and Control" (SSDC) covers both new developments and advances, as well as the state of the art, in the various areas of broadly perceived systems, decision making and control–quickly, up to date and with a high quality. The intent is to cover the theory, applications, and perspectives on the state of the art and future developments relevant to systems, decision making, control, complex processes and related areas, as embedded in the fields of engineering, computer science, physics, economics, social and life sciences, as well as the paradigms and methodologies behind them. The series contains monographs, textbooks, lecture notes and edited volumes in systems, decision making and control spanning the areas of Cyber-Physical Systems, Autonomous Systems, Sensor Networks, Control Systems, Energy Systems, Automotive Systems, Biological Systems, Vehicular Networking and Connected Vehicles, Aerospace Systems, Automation, Manufacturing, Smart Grids, Nonlinear Systems, Power Systems, Robotics, Social Systems, Economic Systems and other. Of particular value to both the contributors and the readership are the short publication timeframe and the worldwide distribution and exposure which enable both a wide and rapid dissemination of research output.

Indexed by SCOPUS, DBLP, WTI Frankfurt eG, zbMATH, SCImago.

All books published in the series are submitted for consideration in Web of Science.

Jader Lugon Junior · Wagner Telles ·
Pedro Paulo Rodrigues · Antônio José Silva Neto

Inverse Problems in the Environmental Modelling of Rivers and Estuaries

Springer

Jader Lugon Junior
Vocational Doctoral Program in Modeling and Technology for the Environment Applied to Water Resources
Fluminense Federal Institute
Campos dos Goytacazes, Brazil

Wagner Telles
Department of Exact, Biological and Earth Sciences of the Northwest Fluminense Institute of Higher Education
Fluminense Federal University
Santo Antônio de Pádua, Brazil

Pedro Paulo Rodrigues
Graduate Program in Computational Modelling of the Polytechnic Institute
Rio de Janeiro State University
Nova Friburgo, Brazil

Antônio José Silva Neto
Department of Mechanical Engineering of the Polytechnic Institute
Rio de Janeiro State University
Nova Friburgo, Brazil

ISSN 2198-4182 ISSN 2198-4190 (electronic)
Studies in Systems, Decision and Control
ISBN 978-3-032-09936-5 ISBN 978-3-032-09937-2 (eBook)
https://doi.org/10.1007/978-3-032-09937-2

This Springer imprint is published by the registered company Springer Nature Switzerland AG
The registered company address is: Gewerbestrasse 11, 6330 Cham, Switzerland

To Lair, Jader (in memoriam), Ângela, and Clarisse
Jader Lugon Junior

To Juliana
Pedro Paulo Gomes Watts Rodrigues

To Melicia, José Roberto and Vera Clice
Wagner Rambaldi Telles

To Gilsineida, Lucas, and Luísa
Antônio José da Silva Neto

Environmental Sciences Foreword

It is a privilege to write the foreword for the book *Inverse Problems in the Environmental Modelling of Rivers and Estuaries*, authored by our esteemed colleagues Jader Lugon Junior, Pedro Paulo Gomes Watts Rodrigues, Wagner Rambaldi Telles, and Antônio José Silva Neto.

This book, organized into seven chapters, examines inverse problems in environmental modeling of rivers and estuaries. This field is essential for understanding and predicting these complex aquatic systems' physical, chemical, and biological processes. This approach supports diagnoses and future scenario assessments, guiding public policies and engineering projects that address anthropogenic impacts and environmental dynamics.

It aligns with a major interdisciplinary trend in environmental sciences in Brazil and worldwide: the technological axis integrating knowledge from applied social sciences, exact sciences, and natural sciences. The authors connect this knowledge with natural sciences through a rigorous mathematical approach, offering critical support for water resource management. The content is highly technical, yet its applications are inherently interdisciplinary.

In Brazil, environmental sciences were consolidated within graduate programs. They were institutionalized by the Coordination for the Improvement of Higher Education Personnel (CAPES) and the National Council for Scientific and Technological Development (CNPq). Currently, the Environmental Sciences area at Capes, which we coordinate, comprises nearly 150 graduate programs, almost half offering doctorates and a third rated at excellence. These programs are not isolated. They interact with areas such as biology, engineering, and applied social sciences, forming a broad, interdisciplinary field of knowledge. Environmental Sciences is not a discipline or a department that competes with others, but a convergence field. The knowledge produced in environmental sciences results from the union of different competencies. Still, these competencies are not brought together for their own convenience or interests, but rather because of the necessity brought about by complex problems that demand them.

It is worth noting that, in environmental sciences, the scientific-technological process cannot exist apart from social and environmental emergencies. It is in this context that knowledge and technology acquire value and importance. Due to its empirical, inter- and transdisciplinary nature, the field of environmental sciences represents the interconnection between science, technology, society, and the environment, requiring consideration of their mutual relationships. Therefore, interdisciplinary knowledge in environmental sciences is not limited to basic or applied science; it is inherently both. It requires high-level and complex knowledge on the one hand and dialogue with the real world and its application on the other.

The book by colleagues Lugon, Rodrigues, Telles, and Silva Neto is a good example of applying advanced knowledge to complex problems, supporting management and public policy. Developed in a graduate studies context and in dialogue with multiple disciplines, it contributes to disseminating and deepening the understanding of environmental sciences.

Enjoy your reading!

Brasília, Brazil

Valdir Fernandes
Katia Viana Cavalcante
Carlos Alberto Cioce Sampaio
(CAPES Environmental Sciences Area Coordinators)

Computational Modelling Foreword

The strong industrial development and consequent shift of population from rural to urban areas resulted in anthropogenic pressures over the aquatic ecosystems and very often in the degradation of their environmental state. The poor state of the aquatic (and of the air) pushed the appearance of very active environmentalist movements claiming for rules and investments in environmental protection and recovery. Public Administration responses resulted in the development of new technologies and rules that were very effective on the reduction of point source pollution and on the improvement of surface water quality. Diffuse pollution and climate change are nowadays the main environmental concerns. They have distinct origins and consequences, but they have in common the fact that the public do not see the link between the source of the problem and the consequences. Mathematical modelling, the central focus of this book, builds that connection.

Direct models use equations to foresee the solution of a particular problem, knowing initial and boundary conditions. They are suitable tools to study scenarios, playing with the pressures (boundary conditions) to choose the optimal response to a generalized problem. They are the basis of the so-called Driver, Pressure, State, Impact and Response (DPSIR). However, the speed of the scenarios testing and of model parameters' estimation can be much increased using inverse-problem techniques. In the inverse models, the aim is to find the initial/boundary conditions responsible for a given solution (i.e., a given state). This book merges both techniques, making it a perfect text for people willing to learn about the two methods and the advantage of their combination to get faster solutions.

This book focuses on pollutant transport in rivers and estuaries, that are in the path of between anthropogenic pressures and the sea. The authors describe equations and methods with detail making the book suitable for students but also for readers with different backgrounds, leveling the knowledge and unifying the language along the chapters introducing formulations and resolution methods, to reach a wide range of objectives and problems.

The book illustrates a wide range of numerical methods to solve direct problems described by differential equations. The Finite Difference Method (FDM), the Finite

Volume Method (FVM), the Finite Element Method (FEM), and the Generalized Integral Transform Technique (GITT) are described and illustrated with examples as well as the Artificial Neural Network (ANN) for computational intelligence. The book also includes case studies solved using the open-source software MOHID initiated at the MARETEC research center in Lisbon University, and the commercial software Mathematica from the Wolfram software house. The Markov Chain Monte Carlo Method (MCMC), Genetic Algorithm (GA), Particle Collision Algorithm (PCA), Luus-Jaakola (LJ), Particle Swarm Optimization (PSO), Artificial Neural Network (ANN), Ant Colony Optimization (ACO), and Golden Section method (GSM) are used to find the solutions for inverse problems.

The book is organized into six main chapters followed by a concluding chapter. Chapter 1 addresses basic transport phenomena concepts, showing equations for hydrodynamics and transport processes and an introduction to the methods used to solve direct and inverse problems. The subsequent chapters address applications.

Dispersion of pollutants in rivers is addressed in Chaps. 2 and 3. Chapter 2 presents the case of a two-dimensional plume using the Finite Difference Method (FDM), calibrating diffusivity both manually and training an Artificial Neural Network (ANN) using real experimental data. In Chap. 3 the objective associated with pollutant dispersion is to find the function that represents the contaminants emission time-dependence using the formulation of an inverse problem. The direct problem is solved using the Generalized Integral Transform Technique (GITT) and the Inverse Problem is formulated via Bayesian inference and solved with the Markov Chain Monte Carlo Method (MCMC).

Estimation of parameters for the two-dimensional horizontal transport model in an estuarine region at the east coast of Brazil is addressed in Chap. 4 using computational intelligence strategies and stochastic optimization methods. Pollutants source location and intensity estimation is done in Chap. 5 for a hypothetical conservative pollutant using a Computational Intelligence method.

Hybrid analytical-numerical solutions are addressed in Chap. 6 using a Generalized Integral Transform Technique (GITT) for a two-dimensional transient problem of transport of non-cohesive sediments. The solution of the inverse problem used Bayesian inference to identify the dispersion and friction coefficients and sedimentation velocities. Finally, Chap. 7 discusses the book's contributions and future trends in environmental modelling.

For the comprehensive approach and for the importance of the subject addressed I recommend this book to readers interested in learning (or teaching) environmental modelling using a wide range of state-of-the-art computational tools to solve direct and inverse problems.

Ramiro Neves
MARETEC—Instituto Superior Técnico
University of Lisbon
Lisbon, Portugal

Acknowledgements The authors acknowledge the financial support provided by Fundação Carlos Chagas Filho de Amparo à Pesquisa do Estado do Rio de Janeiro (CNPq), Conselho Nacional de Desenvolvimento Científico e Tecnológico (CNPq), and Coordenação de Aperfeiçoamento de Pessoal de Nível Superior—Brazil (CAPES)—Finance Code 001. Besides the support through the graduate student scholarships, the research grants from such agencies allowed the acquisition and implementation of the High-Performance Computer Escola de Sagres. It must also be emphasized the partnerships established with the SENAI CIMATEC University in Brazil, and the Atlantic International Research Centre (AIR Centre) in Portugal, which has provided a number of opportunities to the researchers and students during the elaboration of the book. A very warm acknowledgement is due to Prof. Gabriel de Pinna Mendez and Prof. Rizzieri Pedruzzi for their critical reading and useful suggestions provided during the preparation of different versions of the manuscript. And finally, a special thanks to Springer Editors Prasanna Kumar Narayanasamy and Mayra Castro for their invaluable editing assistance.

Competing Interests The authors have no competing interests to declare that are relevant to the content of this manuscript.

Contents

Glossary

ACO	Ant Colony Optimization
ADI	Alternate Direction Implicit method
ANN	Artificial Neural Network
BG	Bubnov-Galerkin
CDF	Cumulative Distribution Function
CFD	Computational Fluid Dynamics
CI	Confidence (credibility) Interval
DE	Discretization Error
DNS	Discretized Numerical Solution
DP	Direct Problem
EDS	Exact Discrete Solution
FDM	Finite Difference Method
FEM	Finite Element Method
FVM	Finite Volume Method
GA	Genetic Algorithm
GITT	Generalized Integral Transform Technique
GS	Gaussian Smoothing Priori
GSM	Golden Section Method
IP	Inverse Problem
LJ	Luus-Jaakola Method
MAP	Maximum a Posteriori
MCMC	Markov Chain Monte Carlo Method
MLP	Multilayer Perceptron
MOHID	MOdelagem HIDrodinâmica (Hydrodynamic Modeling) Platform
NPS	Nodal Projection Space
ODE	Ordinary Differential Equation
OFRE	Objective Function Relative Error
PCA	Particle Collision Algorithm
PDE	Partial Differential Equation
PDF	Probability Density Function
PG	Petrov-Galerkin

PSO	Particle Swarm Optimization
psu	Practical Salinity Unit (grams of salt per 1,000 grams of water)
RE	Rounding Error
RMSE	Root Mean Square Error
SIRE	Source Intensity Relative Error
SM	Suspended Material
SWE	Shallow Water Equations
TE	Truncation Error
TV	Total Variation Priori
WRM	Weighted Residual Method

Nomenclature

$\overline{a_j}, \overline{b_j}, \overline{c_j}$ $\overline{a_1}, \overline{b_1}, \overline{c_1}$	Coefficients of the base functions in the FEM
A	River cross–section area [m^2]
A_c	River mean cross–section area [m^2]
B	Average river width section [m], Eq. (2.23)
c_1, c_2	Cognitive coefficients that provide a balance between the individual and group experience in the Particle Swarm Optimization (PSO) algorithm
C	Substance concentration [kg m^{-3} or mg L^{-1}]
C^*	Filtered potential (concentration) in the GITT formulation [kg m^{-3} or mg L^{-1}]
C_m	Mean value in the cross section of the concentration time average [kg m^{-3} or mg L^{-1}]
$\bar{C}$	Concentration time average [kg m^{-3} or mg L^{-1}]
C'	Concentration fluctuation with time [kg m^{-3} or mg L^{-1}]
c'''	Spatial fluctuation in the cross section of the concentration time average C [kg m^{-3} or mg L^{-1}]
$C_{ams}(x, y, t)$	Analytical manufactured solution of the transport equation [kg m^{-3} or mg L^{-1}]
C_a	Reference sediment concentration at the flow bed under equilibrium conditions [kg m^{-3} or mg L^{-1}]
C_b	Base concentration in the analytical manufactured solution [kg m^{-3} or mg L^{-1}], Eq. (5.21)
C_f	Filter for the potential (concentration) in the GITT formulation [kg m^{-3} or mg L^{-1}]

$C_0, C_{0_1}, C_{0_2}, C_{0_3}$	Concentration at the inlet boundary of a river stretch, or concentration of a substance continuous launch [kg m^{-3} or mg L^{-1}]
$C_{0_{1,1}}, C_{0_{1,2}}, C_{0_{2,1}}, C_{0_{2,2}}, C_{0_{3,1}}, C_{0_{3,2}}$	Concentration peak values in pollutant discharges [kg m^{-3} or mg L^{-1}], Eqs. (3.49)–(3.51)
C_{0_i}	Unknown values of the concentration to be estimated with the solution of the inverse problem [kg m^{-3} or mg L^{-1}]
$C_0^{rec}, C_{0_1}^{rec}, C_{0_2}^{rec}, C_{0_3}^{rec}$	Recovered concentration with the solution of the inverse problem [kg m^{-3} or mg L^{-1}]
C_1	Initial concentration [kg m^{-3} or mg L^{-1}]
C_D	Prescribed concentration at the boundaries in which Dirichlet boundary conditions hold [kg m^{-3} or mg L^{-1}]
$C_e(x, y, t)$	Function representing the concentration in the element *e*, in the FEM
C_{ini}	Concentration initial condition [kg m^{-3} or mg L^{-1}]
$C_j(t)$	Function representing the concentration time dependence in the node *j*, in the FEM
C_{salt}	Total salt concentration [kg m^{-3} or mg L^{-1}]
C_P, C_N, C_S, C_E, C_W	Concentrations at the nodes of the finite volume discretization of the domain in the FVM [kg m^{-3} or mg L^{-1}]
C_n, C_s, C_e, C_w	Concentrations at the boundary surfaces of the finite volume discretization of the domain in the FVM [kg m^{-3} or mg L^{-1}]
CF	Correction factor used for the equalization of the masses of pollutant in different discharge scenarios, Eq. (3.47)
$\tilde{C}_i$	Transformed potential (concentration) in the GITT formulation, Eq. (3.30)
d	Average depth [m], Eq. (2.7)
d_i	Input signals to the neuron in the Artificial Neural Network (ANN) algorithm
d_s	Diameter of a spherical particle [m]
D	Molecular diffusion coefficient [m^2 s^{-1}]
D_x, D_y, D_z	Diffusion coefficients in the directions *x*, *y*, and *z* respectively [m^2 s^{-1}]
D_l, D_t	Longitudinal and transversal dispersion coefficients [m^2 s^{-1}]
D^*, D_x^*, D_y^*	Diffusion coefficient [m^2 s^{-1}]

D^*	Matrix with the turbulent diffusion coefficients [$m^2\ s^{-1}$]
D_{xx}, D_{xy} D_{yx}, D_{yy}	Elements of the turbulent diffusion coefficients matrix [$m^2\ s^{-1}$]
$\hat{D}^*$	Estimated value for the turbulent diffusion coefficient [$m^2\ s^{-1}$]
D_{ij}	Auxiliary term in the transformed potential problem in the GITT formulation
e	Element in the FEM
$E_{S_l}(t)$	Error in the salinity estimation at the time instant t [%]
E_S	Dimensionless constant coefficient in Eqs. (5.26) and (5.27) [$kg\ m^{-3}/kgm^{-3}$]
E_{max}	Value of the maximum error on the salinity
E_r	Dimensionless relative error
f	Coriolis parameter [s^{-1}]
f_N	Prescribed mass flux at the boundaries in which Neumann's boundary conditions hold [$kg\ m^{-2}s^{-1}$]
f_{act}	Activation function in the Artificial Neural Network (ANN) algorithm
F	Vector of residues
F_D	Sediment deposition flux [$kg\ m^{-2}s^{-1}$]
F_E	Erosion flux [$kg\ m^{-2}s^{-1}$]
F_s	Pollutant source intensity [$kg\ m^{-2}s^{-1}$]
g	Gravity acceleration constant [$m\ s^{-2}$]
g_i	Auxiliary term in the transformed potential problem in the GITT formulation
G_i	Auxiliary term in the transformed potential problem in the GITT formulation
$G_{calc\,i}, G_{meas\,i}$	Calculated and measured values, respectively, for the observable variable
$\boldsymbol{G}_{calc}, \boldsymbol{G}_{meas}$	Vectors with the calculated and measured values for the observable variable
h	Water column depth or water mean depth [m]
h_e	Length of the smallest side among all elements in the FEM, Eqs. (5.28) and (5.29) [m]
h^e	Characteristic length of the element e used in the Petrov–Galerkin relation, in the FEM [m]
H	Auxiliary parameter, Eq. (3.23b)
$-H_0$	Estuary bed location [m]
i_p	Index of the position of data acquisition
i_t	Index of the time of data acquisition

$I^e_{1_i}, I^e_{2_i}, I^e_{3_i}, I^e_{4_i}, I^e_{5_i}$	Integrals used in the FEM
J_a	Advective mass flux [kg m^{-2}s^{-1}]
J_d	Diffusive mass flux [kg m^{-2}s^{-1}]
k	Iteration counter or a state index in the MCMC method
K	Diffusion coefficient [m^2 s^{-1}]
K^*	Dimensionless coefficient whose value takes into account the degree of mixing, as well as the number and position of the injection points, Eq. (2.23)
k_s	Rate of substance concentration decay [s^{-1}]
K_x, K_y, K_z	Turbulent diffusion coefficients, Eq. (4.7) [m^2 s^{-1}]
L_1, L_2, L_3, L_s	Pollutant source location
L, L_x	Longitudinal length along a river [m]
L_y	Transversal length across a river [m]
L_z	Vertical length of the domain of interest [m]
$L_{\min}$	Mixing length, Eq. (2.23) [m]
m	Auxiliary parameter used in the determination of the confidence (credibility) interval
M	Pollutant mass [kg]
M_1, M_2, M_3	Total masses of pollutant in different discharge scenarios [kg or g]
M_i	Weighting functions in the FEM
M_2^{CF}, M_3^{CF}	Equalized masses of pollutant in different discharge scenarios [kg or g]
n_p	Total number of positions of data acquisition
n_t	Total number of time instants of data acquisition
$\vec{n}$	Unit outward normal vector
N_p	Total number of unknowns
N_{bs}	Number of boundary surfaces in the FEM
N_e	Number of elements in the FEM
N_d	Total number of experimental data
N_{EP}	Number of estimated points for pollutant source locations
N_{MCMC}	Total number of states in the MCMC method
$N_{burn-in}$	Number of states in the heating phase of the MCMC method
$N_j(x, y)$	Base functions associated to node j, in the FEM
N_n	Number of nodes in the FEM
N_t	Number of time steps
N_n^e	Number of nodes in element e, in the FEM
N_i	Normalized integral in the GITT formulation

N_{tr}	Truncation order in the GITT formulation
o	Output of a neuron in the Artificial Neural Network (ANN) algorithm
p	Confidence level
P	Pressure [N m^{-2} or Pa]
P_j	Unknown to be determined with the solution of the inverse problem
P_{atm}	Atmospheric pressure [N m^{-2} or Pa]
$P(x, t)$	Source term in the filtered potential (concentration) problem in the GITT formulation [kg m^{-3}s^{-1}]
Pe	Peclet number
$\boldsymbol{P}$	Vector of unknowns
P^*	Solution for the inverse problem or solution candidate generated from a given current state P^k in the MCMC method
P^k	Current state in the MCMC method
$\boldsymbol{P}_{exact}$	Vector of known parameters, Eq. (1.27)
$\vec{p}_{i,j}, \vec{p}_{i,j+1}$	Position of the particle i at the time step j and $j + 1$ respectively, in the Particle Swarm Optimization (PSO) algorithm
$\vec{p}_{best_{i,j}}$	Best position of the particle i up to the time step j in the Particle Swarm Optimization (PSO) algorithm
$\vec{p}_{g_j}$	Best position of the whole group of particles up to the time step j in the Particle Swarm Optimization (PSO) algorithm
q	Auxiliary probability density function in the MCMC method
Q	Volumetric flow rate [m^3 s^{-1}]
r, r_i, r_1, r_2	Random numbers
R	Height of the bottom roughness [m]
R_c	Substance sink term [kg m^{-3} s^{-1}]
R_P	Pearson correlation
R_ϕ	Sink term of property ϕ
$\hat{R}$	Estimated value for the height of the bottom roughness [m]
S	Set of solution candidates for the inverse problem, Eq. (3.41)
$S(\boldsymbol{P})$	Sum of squared residuals, Eq. (1.23)
S_1, S_2, S_3, S_4, S_5	Data acquisition sensor location
S_b	Objective function value for the best estimate of the pollutant sources
S_c	Substance source term [kg m^{-3}s^{-1}]
Sc_t	Turbulent Schmidt number

S_0	Declivity [m m^{-1}], Eq. (2.8)
$S_{l,calc}(t)$, $S_{l,exp}(t)$	Calculated and experimental values for the salinity at the time instant t
S_x, S_y	Body force terms per unit mass in the x and y directions [m s^{-2}]
S_ϕ	Source term of property ϕ
t	Time [s]
t_1, t_2	Time instant for the Gaussian shaped pollutants discharge definition, Eq. (3.50) [s]
t_i	Time instant in the discretized time domain [s]
t_0, t_f	Initial and final time of simulation [s]
u^*	Friction velocity [m s^{-1}], Eqs. (2.7), (2.8) and (6.2)
u''	Spatial fluctuation in the cross section of the time average of the velocity component in the x direction $\underline{u}$ [m s^{-1}]
$\bar{u}$	Time average of the velocity component in the x direction [m s^{-1}]
$u''C$	Spatial mean value of the product of the spatial fluctuations of $\bar{C}$ and $\bar{u}$ [kg m^{-2} s^{-1}]
u, u_m	Mean value in the cross section of the time average of the velocity component in the x direction [m s^{-1}]
u, v, w	Components of the velocity vector in the x, y and z directions [m s^{-1}]
u', v', w'	Fluctuation with time of the velocity components in the x, y and z directions [m s^{-1}]
U	Uniform distribution
$\bar{U}$	Mean velocity [m s^{-1}]
$U(z)$	Velocity in the x direction (longitudinal velocity) with dependence on the vertical coordinate z [m s^{-1}]
$\vec{V}$	Total mass flux [kg m^{-2}s^{-1}]
$\vec{v}$	Velocity [m s^{-1}]
$\lvert\vec{v}\rvert$	Modulus of the velocity vector [m s^{-1}]
$\overrightarrow{vel}_{i,j}$, $\overrightarrow{vel}_{i,j+1}$	Velocity of the particle i at the time steps j and $j+1$, respectively, in the Particle Swarm Optimization (PSO) algorithm
$\underline{w}$	Coefficient of inertia in the Particle Swarm Optimization (PSO) algorithm
W	Measurement errors covariance matrix
w_i	Weights representing the synapses of a neuron in the Artificial Neural Network (ANN) algorithm

W_s	Sediment fall (settling) velocity [m s^{-1}]
X	Modified (scaled) sensitivity coefficient
$\bar{X}_{L_S}^{ip}, \bar{X}_{L_S}^{ip}$	Arithmetic mean values of the sensitivity coefficients
x,y,z	Cartesian coordinates [m]
x_i	Representation of the spatial coordinates x and y, for $i = 1$ and 2, respectively
$\boldsymbol{Y}$	Vector with experimental measurements
$\boldsymbol{Y}_{calc}$	Calculated values with the solution of the direct problem, Eq. (3.38)
y_i	Experimental measurements
z_0	Distance from the bottom at which the velocity given by the law of the wall goes to zero [m], Eq. (6.2)

Greek Letters

α	Regularization parameter
β	Constant used in the Petrov–Galerkin relation
β_1, β_2	Constants in the analytical manufactured solution $[(m^5\ kg^{-1}\ s^{-1})^{\frac{1}{2}}]$, Eq. (5.21)
β_i	Eigenvalue in the GITT formulation
ΔC	Variation of the concentration in numerical approximations [kg m^{-3} or mg L^{-1}]
Δx, Δy	Space step in the x and y coordinates, respectively [m]
ΔL_s	Variation in the spatial location in numerical approximations [m]
Δt	Time interval or time step [s]
$\delta(x)$	Dirac delta function [m^{-1}]
ε	Acceptance probability in the MCMC method using the Metropolis–Hastings algorithm
ε_j	Small variation in the value of the unknown P_j, for the computation of the sensitivity coefficients, Eq. (6.15)
$\varepsilon_1, \varepsilon_2, \varepsilon_3, \varepsilon_4$	Small variations in the values of the parameters D_x, D_z, u^* and W_s, respectively, for the computation of the sensitivity coefficients
ε_x	Longitudinal turbulent diffusion coefficient [$m^2\ s^{-1}$]
ζ	Normal distribution
ϕ	A given property in the transport equation, Eq. (2.1)
Γ	Auxiliary parameter, Eq. (3.23a)
γ	Parameter related to the degree of smoothness in the pollutants pulse discharge, Eqs. (3.49) and (3.51)

κ^*	von Karman constant in the logarithmic law of the wall
λ	First order rate coefficient, Eq. (3.18c) [s^{-1}]
μ	Water dynamic viscosity [$kg\ m^{-1}\ s^{-1} = Pa\ s$]
η	Water free surface elevation [m]
θ	Coefficient for explicit or implicit formulation, Eq. (2.15)
σ	Standard deviation
σ_{exp}	Standard deviation of the experimental measurement error
$\sigma_{2,1}, \sigma_{2,2}, \sigma_{3,1}$	Standard deviation in the Gaussian shaped pollutants discharges, Eqs. (3.50) and (3.51) [$kg\ m^{-3}$ or $mg\ L^{-1}$]
σ_q	Transition step, see MCMC Method in Sect. 3.3.2
ν	Kinematic molecular viscosity ($1.3 \times 10^{-6}\ m^2 s^{-1}$)
ν_H	Turbulent viscosity in the horizontal direction [$m^2\ s^{-1}$]
ν_V	Turbulent viscosity in the vertical direction [$m^2\ s^{-1}$]
ν'_H	Horizontal eddy diffusivity [$m^2\ s^{-1}$]
ν'_V	Vertical eddy diffusivity [$m^2\ s^{-1}$]
ν'_C	Molecular diffusivity [$m^2\ s^{-1}$]
ν_{Co}	Courant number, Eq. (2.22)
$\pi_{pos}(\boldsymbol{P}) = \pi(\mathbf{P}\vert\boldsymbol{Y})$	Posterior probability density function
$\pi_{pr}(\boldsymbol{P})$	Prior probability density function
$\pi(\mathbf{Y}\vert\boldsymbol{P})$	Likelihood function
$\pi(\boldsymbol{Y})$	Marginal probability density
π_{GS}	Gaussian Smoothing priori
π_{TV}	Total Variation priori
ψ_i	Eigenfunction in the GITT formulation
$\tilde{\psi}_i$	Normalized eigenfunction in the GITT formulation
ρ	Water density [$kg\ m^{-3}$]
ρ_s	Density of a spherical particle [$kg\ m^{-3}$]
ρ_0	Constant density reference [$kg\ m^{-3}$]
ρ'	Deviation from the constant density reference [$kg\ m^{-3}$]
τ	Upper limit of the time period [0, τ] for the continuous launch of a substance [s]
$\tau_1, \tau_2, \tau_3, \tau_4$	Time instants for the pollutants pulse discharge definition, Eqs. (3.49) and (3.51) [s]
τ_{Wx}, τ_{Wy}	Components of the wind shear stress along the x and y axis, respectively [$kg\ m^{-1}\ s^{-2} = N\ m^{-2}$]
τ_{Bx}, τ_{By}	Components of the bottom shear stress along the x and y axes, respectively [$kg\ m^{-1}\ s^{-2} = N\ m^{-2}$]
$\tau_{xx}, \tau_{xy}, \tau_{xz}$ $\tau_{yx}, \tau_{yy}, \tau_{yz}$ $\tau_{zx}, \tau_{zy}, \tau_{zz}$	Reynolds stress tensor components [$kg\ m^{-1}\ s^{-2} = N\ m^{-2}$]
μ	Water dynamic viscosity [$kg\ m^{-1}\ s^{-1} = Pa\ s$]
μ_x, μ_y, μ_z	Turbulent viscosity coefficients [$kg\ m^{-1}\ s^{-1} = Pa\ s$]
$\bar{\mu}$	Mean value
μ_i	Eigenvalue in the GITT formulation

Ω	Study region
$\partial\Omega$	Boundary of the study region
Ω_e	Element physical domain in the FEM
$\partial\Omega_e$	Boundary of the element e in the FEM

Subscripts

calc	Calculated values
D	Identification of points located at the boundaries, in which Dirichlet boundary conditions hold
exact	Exact values
exp	Experimental values
max	Maximum value
N	Identification of points located at the boundaries, in which Neumann boundary conditions hold

Superscripts

T	Transpose

Chapter 1
Mathematical Models for Pollutants Transport in Rivers and Estuaries

In this chapter the mathematical models usually employed to simulate the transport of pollutants in fluvial and estuarine environments are presented and described. The equations used to model these problems, which are the direct problems of interest in this book, are formulated and discussed, with focus on mathematical issues and the computational implementation. The physical aspects that justify simplified versions of such models are discussed, as well as the main parameters that control the pollutants concentration in those environments. In this chapter the inverse problem approach is also presented, which is used in the estimation of parameters that are needed in the solution of the direct problem computational model for the applications of interest. Furthermore, it is also presented the sensitivity analysis approach, that is used to evaluate the viability of performing such estimation. Closing this first chapter, the book organization is presented with the description of each chapter contents and objectives, to be used as a guide by the reader.

1.1 Introduction

Water resources management, involves diagnosis, monitoring and forecasting of scenarios, one must consider climate change and sustainability aspects that are the fundamental challenge today [21, 89, 189]. Historically, evaluation of dissolved and suspended constituents' behavior in rivers, channels, canals and estuaries was initiated with the development of the so-called "black box" models, in which the removal or production of the constituent was evaluated performing a mass balance to a given stretch of river, channel, canal or estuary. Such approach, was based on intensive monitoring. It was inaccurate, labor intensive and expensive. Presently, the behavior

J. Lugon Junior et al., *Inverse Problems in the Environmental Modelling of Rivers and Estuaries*, Studies in Systems, Decision and Control 246,
https://doi.org/10.1007/978-3-032-09937-2_1

of substances in these bodies of water can be estimated using computational mathematical and models requiring less data for results validation and boundary conditions specification.

Models, can be defined as simplified representations of reality, aiming at understanding the phenomena that govern it. For that purpose, it is necessary that a given model exhibits the description of the essential characteristics of the real world, i.e., the real practical problem at hand, so that its behavior reflects partially or totally the behavior of the modeled system. According to Ref. [9], the models can be classified as:

(a) Iconic models (images, icons): reproduce the appearance of the object to be modeled, providing a qualitative description only, that still synthesizes the whole. Examples include maps and diagrams of blocks.
(b) Analog models: they are a physical representation of a real system. They have an average degree of abstraction and are quite used in engineering. Reduced scale airplane models for wind tunnels testing are a typical example.
(c) Mathematical models: Represent reality using equations, empirical or based in basic principles. In these models, the degree of abstraction is maximum, but they are the less expensive and easier to extend to new problems.

Up to present, the solutions of mathematical models, for both direct and inverse problems, rely on computational methods and tools developed for each model. However, with the advancement of computational intelligence, we may even expect to be able to get solutions from data only and to prescind from the usual formulations based on differential or integro-differential equations. For such approach, combining mathematical and computational intelligence models, applied to the solution of problems with high level of complexity, we use the denomination of computational modelling.

Note that mathematical models do not escape to the definition of "model", distinguishing themselves only by the algorithms and/or computational language used in the resolution of the equations representing the phenomena they seek to model. Thus, in this category of models, equations and functions are used to express the set of interactions that come up in a physical system. According to Ref. [158], a mathematical model is constituted by a set of equations that represent, quantitatively, the hypotheses that were used in the construction of the model, aiming to describe a real system. These equations are solved based on some known or predicted values and are usually tested and validated by comparing their results with the measurements made in the real world.

Mathematical models always have parameters and state variables in their formulation. Parameters describe processes not solved explicitly (e.g. friction in a hydrodynamical model or maximum growth rate in a biological model) and maintain their value unchanged throughout the studied process. The values of the state variables depend on the processes being modulates and their values are expected to change while the model is executed [131, 133]. The model equations are the mathematical description of the conceptual model that supports hypotheses. They give a quantitative description of the system, but should not be used as the description of the system [136, 158].

Mathematical models are used in the most diverse fields of science, and in general can be separated in two broad categories: deterministic and stochastic. In the former, laws established with theoretical support are formulated through equations whose solution can provide the behavior of a given variable. Deterministic models are able to predict, a priori, the behavior of the variables they are modeling. Stochastic models—often called random or probabilistic—are built on the fundamental concepts of Probability and Statistics. As such, they can predict the probability—seen as a measure of confidence—of variables within the ranges of variability of values used to fit the probability functions, distinguishing themselves essentially in this aspect from the deterministic models.

The application of mathematical models in water resources management began in the early twentieth century. The initial impulse for this application was motivated by the need to manage domestic wastewater in cities that were initiating a rapid growth. Population growth together with sanitary awareness aroused the interest for finding alternatives to the release of these wastes into water bodies. Collection, transport and deposition networks began to be thought of. As engineering works would obligatorily be an important part of such initiatives, the need for more precise calculations to design them emerged. It is natural, therefore, to make use of mathematical formulations already consecrated in the field of Fluid Mechanics for this purpose. Since then, much progress has been made, with applications no longer restricted to man-made ditches and reservoirs, but also reaching natural ecosystems themselves through considerably more complex models. According to Ref. [7], these models include applications such as quantification of hydrological cycle processes, hydraulic flow processes in rivers [163], water quality analysis in rivers [36, 37], reservoirs and groundwater, coastal seas and in the atmosphere.

Other examples of mathematical modeling application for management of aquatic systems are given in Ref. [52]: (i) Evolution of oil spills [123, 167]; (ii) Search and rescue [58, 98, 99, 101, 165]; (iii) Potential algal blooms; (iv) Particle transport, and dispersal and distribution in water bodies (fish roe and larvae, contaminants, suspended matter, river waters) [126, 135]; (v) Transport of nutrients and suspended matter [121, 166]; (vi) Transport and dilution patterns of potentially polluting discharges [125, 164, 182, 183]; Eutrophication (nutrients and oxygen) [49, 50]; (viii) Production and updating of long time series of the climate of aquatic environments, and variability of ecosystems (especially at lower trophic levels); (ix) "coastal" climate (currents, turbulence, temperature, salinity, waves, wind, air temperature) [100, 190]; (x) Demonstration of the potential status of coastal or regional climates, as well as state and ecosystem variability, against hypothetical future climate scenarios; and (xi) Quantification of the effects of different management scenarios in coastal environments (namely on the level of primary production, eutrophication, and nutrient transport) [136].

It is important to emphasize that the development of the mathematical modeling of water resources systems was only strengthened from the 1960s onwards, when the advent and increase of computational power and availability of computers began. As described in Ref. [7], the expansion of informatics enabled a revolution in descriptive mathematical techniques (simulation), and normative or prescriptive (optimization) techniques [128].

Mathematical models applied in water resources address both water quantity and quality aspects. Stream flow, velocity fields, water availability must consider water quantity, while maximum pollutant load dilution, plume behavior or removal rates are sometimes used water quality for scenario ranking. However, the water quality aspect will always depend on the water quantity, since the evaluation of the water body quality passes through the present concentration of certain chemical compounds, which can only be defined by the volume turnover rate of the receiving body itself.

When it is intended to simulate the behavior of substances present in a water body, three types of models must be built sequentially. A first model "Hydrodynamic Model" will describe the water body hydrodynamics. The transport will use the results of the hydrodynamic model (velocities and water level) top compute the advective and diffusive transports. They are both physical and quantitative. Finally, a reaction model describing phenomenological processes, such as Chemical and Biological must be applied to simulate the transformation of nonconservative (i.e. reactive) properties. In Fig. 1.1 a schematic representation of these models' interconnections is presented, as well as their inputs and outputs. Some aspects of these models are briefly addressed in subsequent sections.

1.1.1 The Hydrodynamic Model

The first step in constructing a model solution in the field of water resources, is the definition of the spatial domain to be modeled, in order to define the boundary conditions. In the case of a river, these limits are defined by its own banks, its bed and free surface, and by two sections that are arbitrarily set, respectively, as its head and its end. These borders identify the domain of interest but not completely since there are fluxes across them, that must be quantified using the boundary conditions. This becomes clear when it comes to water resources, where a major goal is the computation of the fluxes between the catchment and the river across the banks. Estuaries, understood as regions in which the water of continental origin mixes with marine, are another example of difficult systems since fluxes across the sea boundary are not easy to prescribe due to the oscillation character of the flow.

After domain definition, the first step is to build the hydrodynamic model. Through the boundary conditions, this model will account for all the forcings that induce water movement. In case of reservoirs, the surface wind is usually a major flow forcing, directly through shear stress and indirectly through heat exchange. Wind forcing is usually less relevant for rivers where free surface slope and friction (often confused with the bed slope) are usually the most important mechanisms controlling the velocity. In estuaries tide oscillation and inertia are usually the most important processes. The wide range of processes suggests that the wisest approach is to write

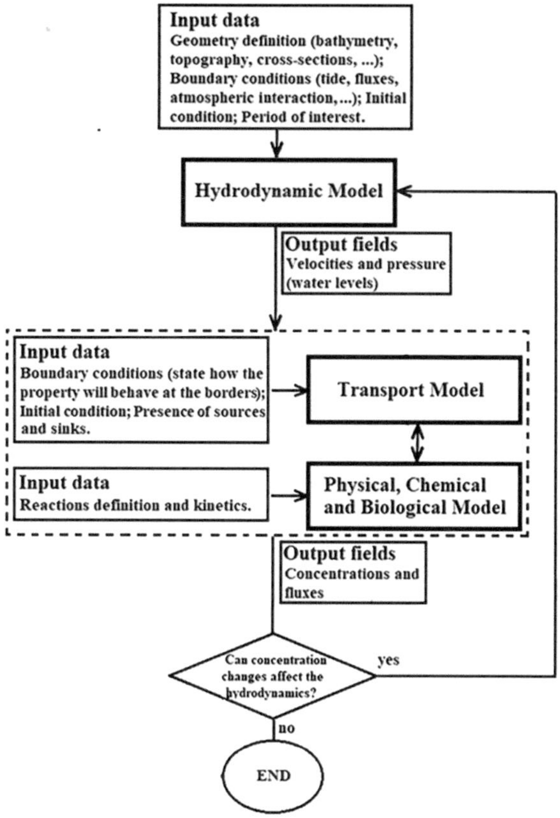

Fig. 1.1 Hydrodynamic, transport, physical, chemical and biological models schematic representation

the equations for a general case, where the hydrodynamic model is understood as a balance of forces and momentum (Second Newton's Law) [17, 143]:

$$\textbf{The Momentum Change Rate} = \textbf{Forces Summation}$$

Or mathematically,

$$\frac{dmu_i}{dt} = \sum F_i \tag{1.1}$$

where m is the mass of the fluid being considered and u_i is its average velocity. F_i are the i components of every force applied on the fluid mass. The product mu_i is the momentum of the fluid mass due to its velocity. Since the mass is constant, the equation above could be written in the well know form: the force is equal to the mass time the acceleration:

$$m\frac{du_i}{dt} = \sum F_i \tag{1.2}$$

In case of incompressible flows, the initial equation can be divided by the volume to obtain,

$$\frac{d\rho u_i}{dt} = f_i \tag{1.3}$$

where f_i hold for the summation of the forces along the i direction (pressure, gravity and friction). Assuming $\rho = const$ the above equation can be written as,

$$\rho\frac{du_i}{dt} = f_i \tag{1.4}$$

where $f_i = pressure\ force + gravity + friction$

The pressure is the pressure gradient, that in case of hydrostatic flows is given by,

$$\frac{Pressure\ Force}{Volume} = -\frac{\partial p}{\partial x_i} = -\rho g\frac{\partial z}{\partial x_i} \tag{1.5}$$

Since in hydrostatic flows the pressure is $p = \rho g z$, where z is the coordinate of the free surface. In shallow water flows the vertical velocity is negligeable and consequently vertical forces are assumed in equilibrium (this is in fact the basic assumption of hydrostatic flows) and consequently gravity force is irrelevant in the equation. The net friction force is the gradient of the shear stress given by $\frac{\partial \tau_{ji}}{\partial x_j}$

As for the pressure forces, the source of movement is not the pressure but its gradient, i.e. the difference between the pressure on the "back" side of the fluid, minus the pressure on the front. In case of friction, the important is not the friction force itself, but the difference between the friction pulling forward on one side of the fluid minus the force pulling backward on the other side.

In case of incompressible flows, the friction forces budget can be represented by

$$\frac{\partial \tau_{ji}}{\partial x_j} = \frac{\partial}{\partial x_j}\left(\mu\frac{\partial u_i}{\partial x_j}\right) \tag{1.6}$$

where μ is the viscosity, molecular in case of laminar flows or turbulent in case of turbulent flows.

Summing up one get the so shallow water equations

$$\frac{du_i}{dt} = -g\frac{\partial z}{\partial x_i} + \frac{\partial}{\partial x_j}\left(\vartheta\frac{\partial u_i}{\partial x_j}\right) \tag{1.7}$$

where $\vartheta = \frac{\mu}{\rho}$ is the kinematic viscosity. Replacing the total derivative by the local plus the convective derivative, we get

$$\left(\frac{\partial u_i}{dt} + u_j \frac{\partial u_i}{\partial x_j}\right) = -g\frac{\partial z}{\partial x_i} + \frac{\partial}{\partial x_j}\left(\vartheta \frac{\partial u_i}{\partial x_j}\right) \tag{1.8}$$

To obtain this equation we have used the Newton's law applied to a material system[1] to relate forces and momentum rate of change, the hypotheses of incompressible and hydrostatic flow and the Newton law of viscosity where the shear stress (friction) is assumed to be proportional to the velocity gradient. The water is effectively quite incompressible and thus this is a minor assumption. The hydrostatic approach assumes that gravity force is balanced by the vertical pressure gradient and consequently there is no vertical acceleration and no need to solve the momentum equation to get the vertical velocity. Below we will show how to compute it from the mass conservation equation.

This equation could have been obtained performing a momentum and force budget to a fixed control volume. This would be called a Eulerian formulation while the one we used, is the Lagrangian formulation. In the Lagrangian formulation we get the total derivative that holds for a material system in the Eulerian formulation we get a local time derivative, local acceleration and an advective acceleration. Performing the Eulerian budget, it would be clear that the local acceleration corresponds to the rate of momentum accumulation inside the control volume and the advective acceleration corresponds to the momentum flux flowing out of the control volume minus the momentum flowing into the volume. When the local velocity increases in time we have an acceleration, the local acceleration, that exits in non-steady flows and when the fluid leaves the volume with a velocity higher than the entrance velocity the material system is accelerated. This happens for many reasons; one clearly happens in areas of cross-section contraction. Upstream of the contraction there is a positive advective acceleration and downstream there is a deceleration. Whatever is the type of acceleration the cause is always the unbalanced forces, i.e. a non-null force summation.

In this expression, the balance of forces per unit mass, represented by the terms on the right side, results in acceleration, expressed by the two terms on the left side. Before representing each of the terms of this expression by its analogues in mathematical notation, a brief description of its nature is appropriate.

The "*advective acceleration*" expresses the velocity increment resulting from the geometric conformation of the simulated domain. Physically it can be observed, for example, in narrowing in rivers and channels, in which it is clearly perceived that the speed of the current increases, when compared to the one observed in broader stretches.

The "*local acceleration*" expresses the rate of increase of speed per unit of time, being the result of the interaction of the other forces.

---

[1] A material system is a system that we can identify. It could be the body of the reader, a car or a certain mass of water, but we should follow always the same molecules. In the fluid mechanics jargon is called the Lagrangian formulation, on opposition to the Eulerian formulation where the control volume is a volume not moving with the fluid and consequently there is fluid moving in and out of the volume due to the relative velocity. There is advective transport.

The "*dispersive mechanisms*" can be understood as the velocity loss due to gradients (differences) in the actual velocity field present in the medium. For example, the water column dynamics of a reservoir that is preponderantly driven by a regime of winds will have markedly vertical gradient of velocities.

The "*pressure gradient*" represents the movement due to level differences between two contiguous water bodies. One way to visualize the effect of this forcing is to imagine a full and hermetically sealed water box. If a single outlet is opened in its bottom, the water will necessarily flow out there, because the mass of water (and the pressure resulting from it) contained in the box is much larger than the one on the outside.

The term "*sources*" brings together all the forces not identified in the other terms. As an example, one has the wind blowing on the surface or the discharge of some affluent to the hydric body.

Finally, "*sinks*" would be all the mechanisms that can dissipate energy (and speed) of the water medium, being clear the example of the friction promoted by the contact of the water mass with the solid boundaries that delimit its domain. The mathematical formulation usually used in the hydrodynamic model is [11, 173]:

$$\begin{aligned}\frac{\partial u}{\partial t}+\frac{\partial(uu)}{\partial x}+\frac{\partial(uv)}{\partial y}+\frac{\partial(uw)}{\partial z}&=fv-\frac{1}{\rho}\frac{\partial P}{\partial x}+\frac{\partial}{\partial x}\left((\nu_H+\nu)\frac{\partial u}{\partial x}\right)\\&+\frac{\partial}{\partial y}\left((\nu_H+\nu)\frac{\partial u}{\partial y}\right)+\frac{\partial}{\partial z}\left((\nu_V+\nu)\frac{\partial u}{\partial z}\right)+S_x\end{aligned}\tag{1.9}$$

$$\begin{aligned}\frac{\partial v}{\partial t}+\frac{\partial(vu)}{\partial x}+\frac{\partial(vv)}{\partial y}+\frac{\partial(vw)}{\partial z}&=-fu-\frac{1}{\rho}\frac{\partial P}{\partial y}+\frac{\partial}{\partial x}\left((\nu_H+\nu)\frac{\partial v}{\partial x}\right)\\&+\frac{\partial}{\partial y}\left((\nu_H+\nu)\frac{\partial v}{\partial y}\right)+\frac{\partial}{\partial z}\left((\nu_V+\nu)\frac{\partial v}{\partial z}\right)+S_y\end{aligned}\tag{1.10}$$

where u, v and w are the components of the velocity vector in the directions x, y and z respectively, f is the Coriolis parameter, ρ is the density, ν_H and ν_V are the turbulent viscosities in the horizontal and vertical directions, ν is the kinematic molecular viscosity (equal to 1.3×10^{-6} m^2 s^{-1}), P is the pressure, and S_x and S_y are the body force terms per unit mass.

From the second, the third and the fourth terms from Eqs. (1.9) and (1.10)

$$\frac{\partial(uu)}{\partial x}+\frac{\partial(uv)}{\partial y}+\frac{\partial(uw)}{\partial z}=\left[2u\frac{\partial u}{\partial x}\right]+\left[u\frac{\partial v}{\partial y}+v\frac{\partial u}{\partial y}\right]+\left[u\frac{\partial w}{\partial z}+w\frac{\partial u}{\partial z}\right]\tag{1.11a}$$

$$\frac{\partial(vu)}{\partial x}+\frac{\partial(vv)}{\partial y}+\frac{\partial(vw)}{\partial z}=\left[v\frac{\partial u}{\partial x}+u\frac{\partial v}{\partial x}\right]+\left[2v\frac{\partial v}{\partial y}\right]+\left[v\frac{\partial w}{\partial z}+w\frac{\partial v}{\partial z}\right]\tag{1.11b}$$

which results in

$$\frac{\partial (uu)}{\partial x} + \frac{\partial (uv)}{\partial y} + \frac{\partial (uw)}{\partial z} = u\left[\frac{\partial u}{\partial x} + \frac{\partial v}{\partial y} + \frac{\partial w}{\partial z}\right] + \left[u\frac{\partial u}{\partial x} + v\frac{\partial u}{\partial y} + w\frac{\partial u}{\partial z}\right] \tag{1.12a}$$

$$\frac{\partial (vu)}{\partial x} + \frac{\partial (vv)}{\partial y} + \frac{\partial (vw)}{\partial z} = v\left[\frac{\partial u}{\partial x} + \frac{\partial v}{\partial y} + \frac{\partial w}{\partial z}\right] + \left[u\frac{\partial v}{\partial x} + v\frac{\partial v}{\partial y} + w\frac{\partial v}{\partial z}\right] \tag{1.12b}$$

and using the "Continuity Equation" (mass balance equation) for incompressible fluid,

$$\frac{\partial u}{\partial x} + \frac{\partial v}{\partial y} + \frac{\partial w}{\partial z} = 0 \tag{1.13}$$

one arrives at the equivalence of the following terms

$$\frac{\partial (uu)}{\partial x} + \frac{\partial (uv)}{\partial y} + \frac{\partial (uw)}{\partial z} = u\frac{\partial u}{\partial x} + v\frac{\partial u}{\partial y} + w\frac{\partial u}{\partial z} \tag{1.14a}$$

$$\frac{\partial (vu)}{\partial x} + \frac{\partial (vv)}{\partial y} + \frac{\partial (vw)}{\partial z} = u\frac{\partial v}{\partial x} + v\frac{\partial v}{\partial y} + w\frac{\partial v}{\partial z} \tag{1.14b}$$

and, therefore, Eqs. (1.9) and (1.10) can be written as

$$\begin{aligned}\frac{\partial u}{\partial t} + u\frac{\partial u}{\partial x} + v\frac{\partial u}{\partial y} + w\frac{\partial u}{\partial z} - fv = &-\frac{1}{\rho}\frac{\partial P}{\partial x} + \frac{\partial}{\partial x}\left((\nu_H + \nu)\frac{\partial u}{\partial x}\right) \\ &+ \frac{\partial}{\partial y}\left((\nu_H + \nu)\frac{\partial u}{\partial y}\right) + \frac{\partial}{\partial z}\left((\nu_V + \nu)\frac{\partial u}{\partial z}\right) + S_x\end{aligned} \tag{1.15}$$

$$\begin{aligned}\frac{\partial v}{\partial t} + u\frac{\partial v}{\partial x} + v\frac{\partial v}{\partial y} + w\frac{\partial v}{\partial z} + fu = &-\frac{1}{\rho}\frac{\partial P}{\partial y} + \frac{\partial}{\partial x}\left((\nu_H + \nu)\frac{\partial v}{\partial x}\right) \\ &+ \frac{\partial}{\partial y}\left((\nu_H + \nu)\frac{\partial v}{\partial y}\right) + \frac{\partial}{\partial z}\left((\nu_V + \nu)\frac{\partial v}{\partial z}\right) + S_y\end{aligned} \tag{1.16}$$

The first term on the left hand side is the temporal evolution of velocities, and the other three terms are the advective transport balance. The first term on the right hand side is the Coriolis force per unit mass, the second term is the pressure gradient, the next three terms are the turbulent diffusion, and the last term represents the body force terms per unit mass.

Assuming hydrostatic pressure, the vertical momentum equation becomes an equation for pressure

$$\frac{\partial P}{\partial z} + \rho g = 0 \tag{1.17}$$

Observe that the vertical axis, z, is positioned in the opposite direction of the gravity acceleration.

The vertical velocity can be calculated assuming an incompressibility hypothesis and constant density according to the Boussinesq approach, expressed by the "Continuity Equation" (mass balance equation), Eq. (1.13).

Integrating Eq. (1.13) from the bottom, i.e. $z = -h$, up to any particular depth z, where the velocity w needs to be calculated, results

$$w(z) = -\frac{\partial}{\partial x}\left(\int_{-h}^{z} udz\right) - \frac{\partial}{\partial y}\left(\int_{-h}^{z} vdz\right) \tag{1.18}$$

Finally, derived from Eq. (1.18), an equation to compute the free surface level can be obtained by integrating over the whole water column, from the bottom $-h$, up to the free surface elevation $\eta(x, y)$,

$$\frac{\partial \eta}{\partial t} = -\frac{\partial}{\partial x}\left(\int_{-h}^{\eta} udz\right) - \frac{\partial}{\partial y}\left(\int_{-h}^{\eta} vdz\right) \tag{1.19}$$

1.1.2 The Transport Model

Once the hydrodynamic model is defined, the next step is to construct the transport model, which can also be represented by an expression whose terms describe the processes responsible for the mass transport of a given substance present in the water body. Put in words, it can be stated as [17, 143]:

Rate of change over time (local) + **Advective transport** =
Dispersive Mechanisms + **External Contributions** + **Sources** – **Sinks**

The first term on the left side of this expression represents the time variation of the concentration of the simulated substance. The second term represents the mass transport of this substance promoted by the velocity field. On the right hand side, as "*dispersive mechanisms*", we have the mass transport due to the gradient (differences in space) of concentrations observed in the medium. "*External Contributions*" would come from other water bodies (e.g. an affluent river), atmospheric deposition (wet and dry) and anthropogenic (point or diffuse) discharges. *Sources* would be all processes and internal reactions (of a physical, chemical or biological nature) that could result in the production of the simulated substance, whereas *sinks* would be processes and reactions that result in the consumption of this substance. The mathematical formulation usually used in the transport model is [11, 143, 145]

$$\frac{\partial C}{\partial t}+\frac{\partial(uC)}{\partial x}+\frac{\partial(vC)}{\partial y}+\frac{\partial(wC)}{\partial z}=\frac{\partial}{\partial x}\left(\nu'_H\frac{\partial C}{\partial x}\right)+\frac{\partial}{\partial y}\left(\nu'_H\frac{\partial C}{\partial y}\right)+\frac{\partial}{\partial z}\left((\nu'_V+\nu'_C)\frac{\partial C}{\partial z}\right)+S_C-R_C \quad (1.20)$$

where C represents the volumetric concentration of the substance of interest, ν'_H and ν'_V are the horizontal and the vertical eddy diffusivity, respectively, ν'_C is the molecular diffusivity, and S_C and R_C are the source and sink terms.

If a model could describe the substance transport with sufficient detail in time and space, one would say that there is no diffusion, everything would be described as advection, no random or turbulent component would happen and, therefore there would be no need for diffusion formulation [145]. Since such detailed model is impossible to build for any practical purpose, and there is always some random component, we name and consider them as the turbulent diffusion effect and, therefore we have the horizontal and vertical eddy diffusivities ν'_H and ν'_V.

In summary, horizontal and vertical eddy turbulent diffusion are caused by velocity gradients that are not calculated in the scale of the model [143], and usually $\nu'_C \ll \nu'_V \ll \nu'_H$.

We must point out that in the following chapters of this book we are also representing the diffusion and dispersion coefficients using the mathematical symbols D and K.

1.1.3 The Physical, Chemical and Biological Models

As previously pointed out, the substance of interest can also be subjected to different phenomenological processes, such as physical phenomena, chemical reactions and biological transformations. Then, additional sets of equations can be adopted to formulate those processes.

Infiltration, sedimentation and evaporation are examples of physical phenomena that can be formulated and incorporated. For example, the sedimentation process can be modelled considering a spherical particle, with diameter d_s, falling down in the liquid column with laminar flow. The velocity of sedimentation W_s obtained using Stokes law is [32, 123, 143, 175]

$$W_s=\frac{g}{18}(\rho_s-\rho)\frac{d_s^2}{\mu} \quad (1.21)$$

where g is the gravity acceleration, ρ_s is the density of the spherical particle, ρ is the water density, and μ is the water dynamic viscosity.

Then, the sedimentation rate for a given substance can be understood as a sink that diminishes the concentration C of a particulate pollution, for instance, and can be modelled as [30, 143]

$$R_C = k_s C = \frac{W_s}{h} C \tag{1.22}$$

where k_s is the rate of substance concentration decay and h is the depth of water at each position, or the mean depth in a particular position inside the domain of interest. One must observe the presence of negative signal in the sedimentation term R_C in Eq. (1.20).

1.2 The Modelling Solution and Implementation (Direct Problem)

The solution of Eqs. (1.13, 1.15–1.19) provides the distribution in time and space of the quantities to be simulated, namely, velocity, water level, concentration of substances, etc. This solution, however, will only be implemented if boundary and initial conditions are properly set in the model. As a boundary condition, we mean the value of the quantities or their fluxes at the boundaries that delimit the simulated domain, whereas initial conditions can be arbitrated or measured, expressing the spatial distribution of the quantities in the domain, before starting the simulation itself. The quality of these information, i.e. initial and boundary conditions, is determinant for the quality of the simulation. It is common sense among modelers that the simulation will never be more accurate than the data prescribed to the model.

The set of equations and the initial and boundary conditions close the mathematical model. However, given the complexity of these equations, in many cases an analytical solution cannot be obtained. That is, even with the closing of the mathematical problem, it is not possible to obtain functions that relate the quantities of interest, i.e. the dependent variables, to the so-called independent variables, which are the time and the three spatial coordinates. Such difficulty leads to the need that numerical solutions be sought, generating what is called numerical simulations. The underlying idea of numerical simulation is the discretization process, which reduces the physical problem, continuous and with an infinite number of variables, to a discrete problem with a finite number of variables, so that it can be computationally solved [194]. Consider, for example, the simulation of the distribution of the concentration of a given substance thrown into a river. If analytical solutions were possible, once the mathematical problem was solved, there would be at any point in the domain a curve that would continuously express the evolution of that concentration over time. However, in the impossibility of obtaining this analytical solution, the numerical resolution will provide, at discrete points, concentration values at also discrete times.

These equations contemplate the simulation of water bodies (both flow and transport of substances) in their highest degree of complexity. However, depending on the peculiarities of the simulated water body, approximations can be assumed, with omission of processes and forcing terms, without loss of simulation quality, but with a reduced computational effort, time, and cost. More importantly, two and one-dimensional approaches can be implemented, as well as non-transient (time invariant) models can also be applied. For example, narrow rivers can often be simulated through a one-dimensional approach, whereas shallow lagoons can be satisfactorily modeled using a two-dimensional approach.

Regardless the adopted simplifications, such models must be able to describe accurately the hydrodynamics and transport of solutes or suspensions in the simulated water body. Ideally, they are presented as the most appropriate tools for diagnosis and prognosis of aquatic environments, since several scenarios can be generated, simply by modifying some of the inputs of the model. However, before any practical application, these models must be calibrated, when model parameters are carefully altered, so that a better agreement between simulation and observation is obtained. If this agreement is satisfactory up to a certain level that is arbitrated by the developer, the model is said to be calibrated. If this model is able to reproduce another field situation, different from the one used in the calibration process, the model is said to have been validated, and is therefore ready to be applied. Thus, although they are always abstractions from reality, if properly calibrated and validated, these models reduce the time, cost and risk of impact analyzes performed on water bodies.

A computational platform can be seen as an operational modeling system, that is, as an infrastructure of coupled models, a set of input information (such as bathymetry and initial conditions), and an automatic (including remote) monitoring network; all integrated in an operational model, so that the results and information obtained are presented in a clear way to the end users [50].

Despite the modelers efforts to generate simulations that are closer to reality, it is well known that natural processes are much more complex than the models developed, mainly because they involve different time and space scales, and are influenced usually by a large number of parameters, that may even be correlated. Modeling can provide a qualitative understanding of a particular process, but, on the other hand, a global understanding of the dynamics of an environment can be reduced. From the same reasoning, the use of detailed mathematical equations in order to express complex physical processes may become impracticable, which would require the use of simpler models [66, 169].

This unfeasibility can occur, even due to computational limitations, because it is not always possible to represent detailed scales explicitly in the model. In certain cases, these equations can be seen as a set of hypotheses that interconnect inputs and outputs of the system.

There are countless applications of numerical simulation in water quality problems of rivers, channels and estuaries, such as the transport of suspended material (SM), the behavior of industrial effluent plumes in rivers and estuaries, as well as losses or infiltration and contamination in channels [192].

The solution of the model is obtained using the computer code, and it is referred to as the solution of the Direct Problem [118], when the modeler informs the set of model parameters and other information, such as geometry, initial and boundary conditions,

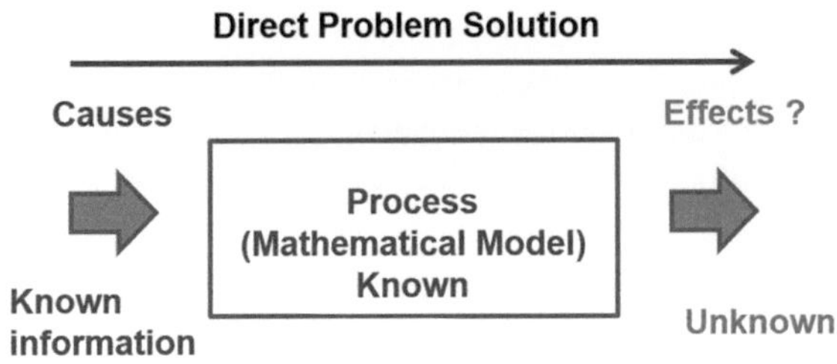

Fig. 1.2 Direct Problem schematic representation [149]

and sources and sinks, i.e. the "Cause" that characterizes the situation of interest and the mathematical equations that are capable of representing the phenomena involved. The "Effect" for that set of equations with the corresponding given information is, therefore, obtained from the Direct Problem solver. The description just presented is schematically represented in Fig. 1.2.

1.3 Inverse Problem

In the equations previously presented, there is a range of parameters that are not always calculated directly. On the other hand, their estimation is often of crucial importance to the simulation accuracy. The definition of the best values for these parameters can be conducted through exhaustive search, or trial and error, which can consume considerable effort and, worse, generate values that, although enabling the simulation for a specific situation, do not allow it to be extended to other situations, when it is said that the model can be calibrated, but not validated. Moreover, such strategy can commonly generate a multiplicity of equally acceptable solutions, from which it is difficult to distinguish the best. Alternatively, this adjustment can be driven by algorithms, among which the inverse problem analysis technique stands out.

Inverse problems have the characteristic of being ill-posed, and several solution techniques have been developed and applied in engineering and other branches of science [118]. In simulations applicable to water resources, a few examples are found in the literature. For example, in Ref. [147] it was used a modification of Newton's method to estimate non-point sources of fecal coliforms in order to establish the maximum allowable load for the Wye River, Maryland/USA. In Ref. [196] it was estimated the salinity at the estuarine marine boundary with the application of inverse problem. In Ref. [174] it was proposed a new algorithm to estimate the open boundary conditions in hydrodynamic models applied to tidal channels.

In the solution of a direct problem, some information may not be known, such as geometry, boundary conditions, initial condition, process parameters and physico-chemical properties. However, assuming that transient experimental data are available, of observable variables that can be measured, one can then try to estimate the unknown quantities.

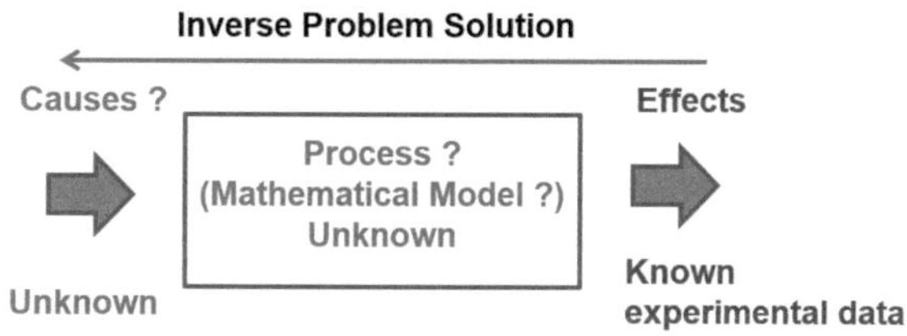

Fig. 1.3 Inverse Problem schematic representation [150]

In Fig. 1.3 the Inverse Problem is schematically represented. With partial information available for the "Effect", one attempts to obtain estimates for the unknown information related to the model developed, such as the process parameters, the domain geometry, initial conditions, and sources and sinks, i.e. the "Cause".

There are several different ways to formulate and solve the inverse problems [15, 25, 118, 122, 141, 150]. Such ways may be classified in two groups [148, 152, 154]: (i) explicit formulations; and (ii) implicit formulations.

In the explicit formulations the equations that represent the direct problem of interest are manipulated, resulting in a mathematical expression in which the unknown parameters to be determined appear explicitly [4–6, 65, 77, 151]. In the resulting expressions the known experimental data is also present. Such approach is usually elegant, yielding insight on the inverse problem at hand, but is very sensitive to the noise present in the experimental data, requiring regularization strategies.

It is very common to formulate the inverse problem as an optimization problem, in which one seeks to minimize the objective function given by the summation of the squared residues between a calculated value for the observable variable and the corresponding experimentally measured value [118]. The calculated quantity is obtained using the direct problem model and estimates for the unknown parameters we seek to determine with the solution of the inverse problem. Therefore, the unknowns appear implicitly in the formulation of the inverse problem.

The just described implicit formulation for the inverse problem is mathematically represented next. Consider the following objective function:

$$\begin{aligned} S(\boldsymbol{P}) &= [G_{calc}(\boldsymbol{P}) - G_{meas}]^T W^{-1} [G_{calc}(\boldsymbol{P}) - G_{meas}] \\ &= F^T W^{-1} F \end{aligned} \tag{1.23}$$

where G_{meas} is the vector containing the measurements, G_{calc} is the vector with the values calculated by the mathematical model for the direct problem, P is the vector of unknowns, i.e., the calibration parameters of the model, W^{-1} is a diagonal matrix with its elements given by the inverse of the variances of the measurement errors, and F is the vector of residues, whose elements are given by

$$F_i = G_{calc_i}(\boldsymbol{P}) - G_{meas_i};\ i = 1, 2, \ldots N_d \tag{1.24}$$

where N_d is the total number of experimental data.

The solution of the inverse problem, P^*, which minimizes the norm given by Eq. (1.23), is therefore sought, i.e.

$$S(\boldsymbol{P}) = S(\boldsymbol{P}^*) \tag{1.25}$$

Therefore, for the solution of the inverse problem implicitly formulated, one seeks to minimize the objective function given by Eq. (1.23). For that purpose, are used deterministic, stochastic or hybrid methods, combining deterministic and stochastic methods, or even stochastic with stochastic methods [3, 59, 86, 93, 94, 103, 104, 110, 118, 149, 150, 153, 154, 159, 168, 177–180, 184, 185].

In recent years it has been observed an increasing interest on using statistical techniques for the solution of inverse problems, such as the Bayesian inference [57, 83, 84, 87, 88, 122, 156, 163]. Such approach is computationally very intensive, but instead of providing just one value for the unknown parameters, what happens with the previously described approaches for the inverse problem solution, yields the probability density function for the sought unknowns, among other relevant statistical quantities.

1.3.1 Sensitivity Analysis and Design of Experiments

Sensitivity analysis plays an important role in many aspects related to the formulation and solution of inverse problems. Such analysis can be developed in order to evaluate the behavior of observable variables with respect to the parameters of interest, which can be carried out, for example, by the modified, or scaled, sensitivity coefficient X, which is formally expressed as [15, 86, 99, 100, 124]

$$X^i_{P_j} = P_j \frac{\partial G_{calc_i}}{\partial P_j}, i = 1, 2, \ldots, N_d, j = 1, 2, \ldots, N_p \tag{1.26}$$

The sensitivity of G_{calc} with respect to the unknowns P, which one wishes to estimate, must be high enough to allow estimation within reasonable confidence limits. Moreover, when two or more parameters are simultaneously estimated, their effects on the observable variable must be independent, otherwise it may be difficult to distinguish their influence, thus generating poor estimates.

The sensitivity analysis is equally important for the design of experiments, that is, the sample planning, involving position in the space and adequate time interval for the data collection that will be used in the calibration process.

In addition, uncertainty analyzes can be conducted with respect to the estimates obtained for the parameters. Among these analyzes tools, the Monte Carlo method has had wide application in water quality models [63]. This method consists in the a priori adoption of a probability density function (PDF) for a certain parameter, that is, a function that relates the possible values assumed by this parameter and the respective probability of occurrence. From the concepts of probability, it is known that the integral of the PDF is the cumulative distribution function (CDF) that represents the cumulative probability up to a certain value assumed by the random variable, in this case the parameter in question. The CDF assumes values from 0 to 1 (or 0 to 100%). If random numbers are repeatedly generated within this range, from the CDF one can obtain corresponding values for the parameter that is estimated and consequently for the variable of interest, which in the study proposed here may be the concentration of the substance whose transport is being simulated, the water level or velocity components. It follows from the procedure that a probability distribution is obtained for the variable that is being simulated.

1.3.2 Experimental Data

It is usual that when modelers study a particular phenomenon, no real experimental data is available for the case of interest, but equations and parameters are, from literature research. So, it is common practice to create synthetic experimental data, using the direct problem solution obtained with such equations and corrupting it with some known statistic distribution random error.

Considering a normal distribution for the measurements error, synthetic experimental data can be generated with

$$G_{meas_i} = C_{calc_i}(P_{exact}) + \sigma_{exp} r_i, \quad i = 1, 2, \ldots, N_d \tag{1.27}$$

where $C_{calc_i}(P_{exact})$ are the concentration obtained from the solution of the direct problem with the vector of known parameters P_{exact}, r_i are random numbers in the range $[-1, 1]$ and σ_{exp} emulates the standard deviation of measurements errors, and N_d, as previously mentioned, is the total number of experimental data.

It is important to understand that experimental error can be originated from many causes, among them: the amount of tracer material, the river velocity, the sampling position, the measurement techniques and instruments, etc.

1.3.3 Brief Description of Methods Used to Solve Inverse Problems

In this section a brief description of the methods used to solve the Inverse Problems of interest in the book is presented [92–96, 103, 110, 118, 124–126, 134, 135, 152, 162, 169, 177–180, 182].

Deterministic methods

An important class of methods available to solve inverse problems is called "deterministic methods", usually gradient-based, and they search for the minimum of the objective function described by Eq. (1.23) looking at the first derivative of the objective function with respect to the parameter of interest. Using such methods, one searches for a point with a zero derivative for each parameter of interest.

This class of methods generally use an iterative procedure, and after a certain number of iterations will hopefully converge to the minimum of the objective function.

The advantage of that class of methods is that they are usually fast when compared to other strategies. On the other hand, they can stop when a local minimum is obtained and the best solution is never found.

Stochastic methods

Stochastic methods do not make use of the gradient of the objective function to descent to the minimum. Some of them tend to mimic some process inspired in the nature to look for the minimum of the objective function, others use other logical strategies in order to narrow the search region and to determinate the best solution.

Stochastic methods can ideally converge to a global minimum, but they are computationally slower than the deterministic methods. In fact, stochastic methods can easily require such a huge number of evaluations of the objective functions that, in some cases, they can become non-practical.

Statistical inference methods

The objective of statistical inversion methods is to reformulate the inverse problem as a problem of statistical inference, and then, attempts are made to incorporate into the model all the available information. The main objective of this approach is to reduce uncertainties about the parameters of interest.

Bayesian inference methods used to solve inverse problems can be described as techniques that do not usually use the inversion of matrix operators. Then, this methodology has a certain advantage in relation to traditional inversion techniques, since such operators are subject to numerical instabilities. On the other hand, they are also computationally intensive.

Hybrid strategies

In order to take advantage of the best characteristic of each method and overcome some of their difficulties, are used the so-called hybrid algorithms. They are developed taking advantage of the robustness of stochastic methods and also the fast convergence of the deterministic methods. Each method offers different degrees of convergence, reliability and robustness, and the hybrid method can also swap between them along the iterative minimization procedure according to some previously programed strategy.

1.4 Chapters Contents

After this first chapter that presents the mathematical formulation, discusses some computational aspects, and introduces the inverse problem approach, in the second chapter a Finite Volume Method is presented and an Artificial Neural Network is trained, in order to simulate the behavior of a plume of pollutants in a fluvial medium. The results are compared with experimental measurements obtained from the release of a conservative substance in a rectilinear stretch of the Macaé River, an important river located in the southeastern region of Brazil. A trial-and-error approach is used for the calibration of the model. A substance is considered conservative when it remains unchanged in its chemical form and does not undergo degradation or transformation in the system of interest.

In the third chapter the estimation of the timewise dependence that can represent the release of a conservative pollutant in a hypothetical stretch of a river is presented. The direct problem is solved by the analytical–numerical hybrid method known as the Generalized Integral Transform Technique (GITT). In contrast, the inverse problem is formulated via Bayesian inference and solved through the Markov Chain Monte Carlo Method (MCMC).

In the fourth chapter it is presented the estimation of both the roughness mean height and turbulent diffusion coefficient used in a 2-D Horizontal transport model, which was applied in the estuarine region of the Macaé River/Brazil. The estimates are obtained within the framework of Computational Intelligence, with the application of stochastic optimization methods, such as Genetic Algorithm (GA), Particle Collision Algorithm (PCA), Luus-Jaakola (LJ), Particle Swarm Optimization (PSO), and Artificial Neural Networks (ANN).

In the next chapter, the fifth one, Inverse Problem techniques are applied to estimate the source parameters (location and intensity) of a hypothetical conservative pollutant released in estuarine waters. The case study considered here is the Macaé River estuary, in Brazil. The pollutant transport was modeled by the advection–diffusion equation. For the source location estimation was used also a Computational Intelligence approach, with the application of the Luus-Jaakola (LJ), Particle Collision Algorithm (PCA) and Ant Colony Optimization (ACO) methods, and to estimate the source intensity was used the Golden Section (GS) method.

In the sixth chapter, Inverse Problems in Non-cohesive Sediment Transport Model are formulated and solved, using again the GITT for the direct problem and Markov Chain Monte Carlo Method (MCMC) with Bayesian inference to solve the inverse problem of interest.

Finally, in the seventh chapter final remarks are presented, that can be drawn from the research results on pollutant transport in rivers and estuaries presented in the previous chapters, with a number of different Direct and Inverse Problems formulation and solution methods. Some future trends are also presented.

Chapter 2
Estimation of Dispersive Parameters in Rivers—Model Calibration

In this chapter we discuss the solution of a computational transport model to simulate the behavior of a plume of pollutants in a fluvial medium. Therefore, advective and dispersive terms are studied considering a two-dimensional problem. The results are confronted with experimental measurements obtained from the release of a conservative substance in a rectilinear stretch of the Macaé River, an important river located in the southeastern region of Brazil. The Direct Problem was solved using the Finite Volume Method (FVM) and an Artificial Neural Network (ANN) was trained using simulated results. A trail-and-error approach is used in the calibration of the model, comparing experimental data and simulation results, being determined the best values of the transversal and longitudinal dispersion coefficients.

2.1 Introduction

One of the greatest challenges of this new millennium is to avoid the scarcity of fresh drinking water in the world. A study presented in July of 2000 issue of the Science journal shows that approximately two billion people, a third of the world's population at that time, was already facing the serious problem of water shortage. Irrigation water may be lacking in several countries, especially the poorest. The continents and regions most affected by the lack of water are Africa, Central Asia and the Middle East. Between 1990 and 1995, the need for freshwater increased about twice as high as the world's population. Global water demand will increase by 20 to 30% by 2050,

Francine de Almeida Kalas, from the Polytechnic Institute of the Rio de Janeiro State University and Cássius Marcelo Dutra Pessanha, from Petrobras—Brazilian Oil S.A., are also co-authors of this chapter.

J. Lugon Junior et al., *Inverse Problems in the Environmental Modelling of Rivers and Estuaries*, Studies in Systems, Decision and Control 246,
https://doi.org/10.1007/978-3-032-09937-2_2

and the global population is expected to have an increment of higher percentage [21]. This increase is due to the disproportionate raise in the high consumption of water in industrial activities and agricultural areas. According to Refs. [89, 189], while agriculture accounts for roughly 70% of freshwater withdrawals, industrial (≈20%) and domestic (≈10%) uses are the main drivers of increasing water demand. But, even with the food production increasement by more than 100% in the last 30 years, the food requirements of a growing global population by 2050 will increase something around 50% (with respect to 2012). Considering that increasing food demands combined with the climate change impacts in the water cycle, the world will need to improve the agricultural water management and also the environmentally sustainable food systems. Water quality is also a great concern because in lower-income countries, poor ambient water quality is mainly due to low levels of wastewater treatment, whereas in higher-income countries, runoff from agriculture poses the most serious problem. Unfortunately, water quality data remain sparse worldwide. It is extremely important to note that only about 2.5% of the water in the planet Earth are freshwater, and approximately 0.08% is in regions accessible to humans [19].

One of the main factors contributing to this possible water scarcity on our planet is the pollution of rivers, lakes and oceans. Such water pollution originates from several sources, among which are the domestic effluents, urban and rural runoff and industrial effluents [29]. The last one can be traced back until the Industrial Revolution in England in the second half of the eighteenth century, which presented patterns of waste generation that had never been seen before, producing quantities that were much greater than the absorption and recycling capacity of nature, with particular emphasis on non-biodegradable synthetic materials.

Even though the environmental contamination resulting from human activities has seen a vertiginous increase, the search for preventive or remedial measures for the problem only gained momentum in recent decades. As a result, the use and improvement of models, which are formulated with physical, biological and chemical principles, along with the corresponding mathematical equations, are becoming more frequent, being able to facilitate the monitoring and environmental assessment, as well as the minimization of water resources pollution and environmental impacts [16].

Great emphasis is placed on the study of the contamination of rivers and estuaries. Contributing to these studies, Ref. [47] presents a work focusing on dispersion in turbulent flows. Later, in Refs. [53, 109] is presented a discussion addressing topics such as the determination of the length needed for a pollutant to homogenize across the cross-section of a river, taking into account its location of release, as well as influences caused by the dispersion coefficients and turbulence in the mixtures occurring in rivers and estuaries. Several techniques have been developed to solve these models. Since many of them have no analytical solution, numerical methods have been adopted.

In the most recent decades several authors have been dedicated to the purpose of creating numerical tools to solve these mathematical models and, consequently, to estimate the behavior of these pollutants along watercourses. We find in the literature a list of published works in this context. Works such as Ref. [42], which uses direct methods to estimate the dispersion coefficient in rivers of th4 city of Ouro Preto—MG/Brazil, Ref. [70] on river flow models, as well as Refs. [136, 187] on the transport of pollutants in lakes and coastal waters applied to environmental engineering and Ref. [112] on the three-dimensional modelling of the estuary of the Sado River in Portugal, are just some of the examples.

Another solution to model the behavior of contaminants in watercourses, that has been growing in great proportion, adopts the Computational Fluid Dynamics (CFD) concepts, which are based on the analysis of systems involving mass, momentum and energy conservation, besides other phenomena associated with them, based on computer simulations. For instance, the authors of Refs. [107, 127, 191] address in their works the formulation, solution techniques and application of CFD in fluid flow and/or heat transfer problems.

A large number of commercial software packages for this purpose are available in the market, and even in free versions on the Internet, such as MOHID (MOdelagem HIDrodinâmica), which was developed by a group of researchers in the University of Lisbon (Superior Technical Institute) located in Portugal, and which has its structure based on the finite volume method.

In this chapter we present the proposition, mathematical formulation and implementation of a numerical solution using the finite volume method for a direct problem of transport of pollutants in rivers. The inverse problem of interest consists in determining the parameters for the application in water quality models, as well as to serve as a tool to assist in the process of decision making concerning the management of water resources.

The model calibration was performed by comparing the simulated data with experimental results obtained in a field work carried out in a section of the Macaé River, located in the State of Rio de Janeiro, in the southeast region of Brazil [182]. Such results can be used in the development of a support tool to estimate the best point of release of industrial waste in the Macaé River by the Thermoelectric Plant Mario Lago Ltda., for instance, since the experimental test was carried out in the region of launch and monitoring of industrial waste by this plant.

2.2 Macaé River Characterization

The Macaé River is located in the State of Rio de Janeiro in Brazil, and it flows from the Nova Friburgo region to the Atlantic Ocean (Fig. 2.1).

Throughout its extension, the Macaé River has undergone through strong interventions in its drainage network, where the impoundments and the rectifications stand out. One of these major changes can be found in the lower reaches of Macaé. This 25 km stretch hardly shows ciliary forest, which would preserve the ravine, retaining moisture. Its deforestation has caused the silting of the river [35].

The main uses of water are: supplying neighbor municipalities, industrial supply for oil, gas and electrical energy production chain, and irrigation. In the basin, critical problems such as the degradation of preservation areas, especially in the

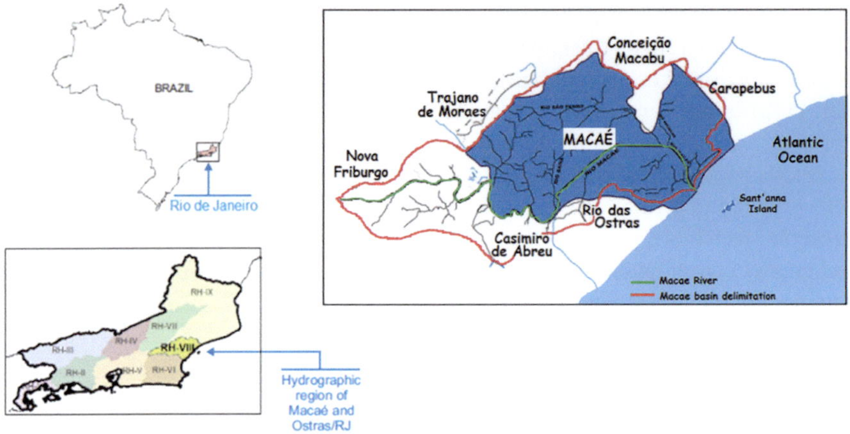

Fig. 2.1 Limits of the Macae River basin [178]

mouth, threatened fauna and flora refuges, sedimentation, mining and the launching of domestic and industrial effluents without treatment.

The study area for which the simulation was developed is located in the lower course of the Macaé River, near the Mário Lago Thermoelectric Power Plant, kilometer 164 of the BR 101 highway, in the municipality of Macaé-RJ Brazil [106, 134, 182], as shown in Fig. 2.2.

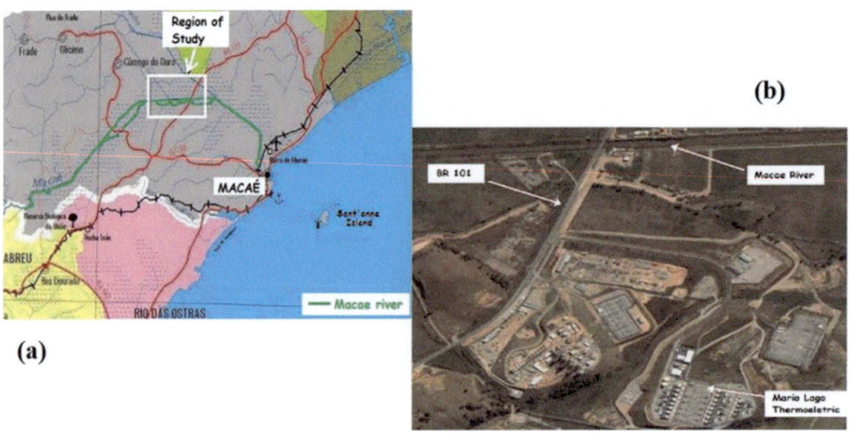

Fig. 2.2 Macaé region of study (**a**) and aerial view of the Mário Lago Thermoelectric Power Plant (**b**)

2.3 Mathematical Formulation and Numerical Solution

The mathematical formulation used in the simulation of the transport of substances in rivers obeys conservative equations, being represented generally by Eq. (1.20). According to Ref. [109], the rate of change in time or the accumulation of a given intrinsic property in the control volume is equal to the sum of the flows or rate of transport over time, across all surfaces of the control volume. The control surfaces are open borders or faces of the control volume. To this variation, any sources or sinks which are present in the medium under analysis are added or subtracted. As in Eq. (1.20), the transport of a given property ϕ with constant density is mathematically formulated as

$$\begin{aligned} &\frac{\partial \phi}{\partial t}+\frac{\partial (u\phi)}{\partial x}+\frac{\partial (v\phi)}{\partial y}+\frac{\partial (w\phi)}{\partial z}= \\ &\frac{\partial}{\partial x}\left(D_x\frac{\partial \phi}{\partial x}\right)+\frac{\partial}{\partial y}\left(D_y\frac{\partial \phi}{\partial y}\right)+\frac{\partial}{\partial z}\left(D_z\frac{\partial \phi}{\partial z}\right)+S_\phi-R_\phi \end{aligned} \quad (2.1)$$

where u, v and w are the components of the velocity vector in the directions x, y and z, respectively, D_x, D_y and D_z are the diffusion coefficients in the x, y and z directions, and S_ϕ and R_ϕ are the source and sink terms of the property ϕ.

2.3.1 Mathematical Model of the Proposed Problem

The simulation discussed in this chapter is focused on the evaluation of the behavior of a conservative pollutant, i.e. $S_\phi = R_\phi = 0$ in Eq. (2.1), released in the region of the lower course of the Macaé River. In this section the river presents a fairly regular geometry, with a practically constant width and uniform cross section, i.e. the course actually resembling a channel. The Macaé River in this region can be classified as vertically well mixed, that is, presents near homogeneity patterns along the water column [7].

As the Macaé River stretch adopted for the simulation has the spatial domain with regular geometry, without curves, and presenting low depths, a two-dimensional approach that retains only the advective term in the longitudinal direction can be used. The coefficients D_l and D_t, previously defined as diffusion coefficients D_x and D_y, are referred to here as longitudinal and transversal dispersion coefficients, respectively [114]. Thus, Eq. (2.1) with $\phi = C$ and constant velocity u can be rewritten as:

$$\frac{\partial C}{\partial t}+u\frac{\partial C}{\partial x}=D_l\frac{\partial^2 C}{\partial x^2}+D_t\frac{\partial^2 C}{\partial y^2};\ 0<x<L_x;\ 0<y<L_y;\ t>0 \quad (2.2)$$

where C is the concentration. With the following boundary and initial conditions:

$$C(0, y, t) = C_0(y, t);\ 0 \le y \le L_y;\ t > 0 \tag{2.3}$$

$$\frac{\partial C(L_x, y, t)}{\partial x} = 0;\ 0 \le y \le L_y;\ t > 0 \tag{2.4}$$

$$\frac{\partial C(x, 0, t)}{\partial y} = \frac{\partial C\left(x, L_y, t\right)}{\partial y};\ 0 \le x \le L_x;\ t > 0 \tag{2.5}$$

$$C(x, y, 0) = C_1(x, y);\ 0 \le x \le L_x,\ 0 \le y \le L_y \tag{2.6}$$

where C_0 is the concentration at the inlet boundary of the river stretch, C_1 is the initial concentration, and L_x and L_y are the longitudinal and transversal lengths of the river stretch, respectively.

The schematic representation of the section of the Macaé River adopted for the simulation is presented in Fig. 2.3.

The solution of Eq. (2.2) requires knowledge of the velocity (u) and of the longitudinal and transverse dispersion coefficients (D_l, D_t). There are some empirical formulations that allow the estimation of these coefficients. For example, the transversal dispersion coefficient (D_t) is expressed as [53]:

$$D_t = 0.15d\ u^* \tag{2.7}$$

where d is the average depth and u^* is the friction velocity.

For the friction velocity u^* the following correlation can be used [109]:

$$u^* = \sqrt{gdS_0} \tag{2.8}$$

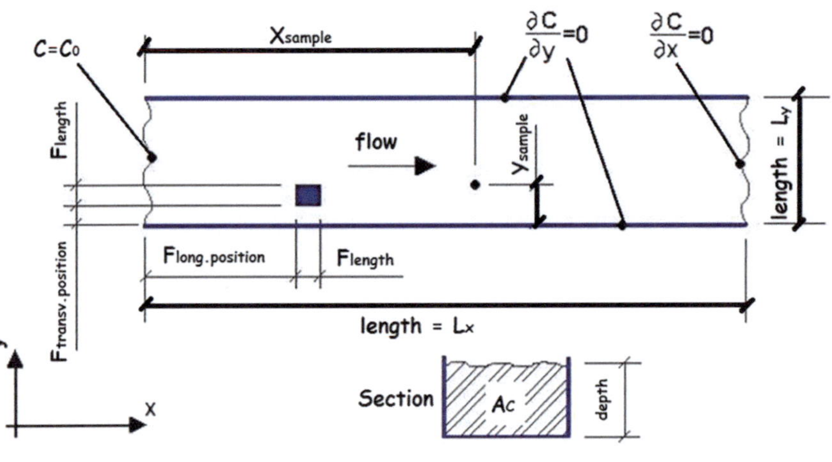

Fig. 2.3 Schematic representation of the Macaé river stretch in which the fieldwork was performed. *Source* Adapted from Ref. [103]

where g is the gravity acceleration and S_0 is the declivity.

Using Eqs. (2.7) and (2.8) it is possible to find an estimate for the transverse dispersion coefficient of the problem proposed here with an approximation error of $\pm 50\%$.

Equations of empirical and semi-empirical nature, generally called practical formulas, have been developed for the prediction of the coefficient of longitudinal dispersion. These equations relate such coefficient with physical quantities of easy attainment or availability, such as mean depth, width, slope and mean velocity or flow rate [42].

In this chapter the longitudinal dispersion coefficient is expressed by the following relation [96]:

$$D_l = 5.93d\, u^* \tag{2.9}$$

where the factor *5.93* is found empirically [47].

As for the longitudinal velocity u, Ref. [109] states that some estimates may be based on the flow rate and a cross-sectional area of the mean historical data, or obtained with a measurement apparatus.

Taking the approach that considers a mean cross-section area, the corresponding mean longitudinal velocity can be calculated by:

$$u = \frac{Q}{A_c} \tag{2.10}$$

where A_c is the river mean cross-section area, and Q is the volumetric flow rate of water.

The model presented [134] is capable of simulating the transport of constituents in solution or in suspension. In the case of an instantaneous pollution discharge in the river, the initial condition will have a specific characteristic according to the mesh adopted in the simulations. Thus, the initial condition $C(x, y, 0) = C_1(x, y)$ will have concentration equal to that verified in the river, except for the cell that will receive the discharge, which will have a variable concentration according to the adopted mesh.

In the impossibility of analytically solving the model that governs the problem described here, a numerical solution is sought, that is, one ceases to work with the continuous physical domain as in the analytical approach, to work with the discrete computational domain, and then a system of equations must be solved by using numerical methods.

According to Ref. [20], analytical methods have the disadvantage of being usually applicable only to problems whose simplifying hypotheses deviate them too much from the real physical phenomena and, generally, can only be applied in simple geometries. However, although analytical solutions do not always represent reality, they should not be discarded, since they are very important for validating numerical models, and their solutions are of the closed type [20].

2.3.2 Numerical Solution

With the fast technological development in recent decades, mainly with regard to digital computers computational power, which generated a significant gain both in resources availability and in speed in data processing, several numerical methods have been developed in order to arrive at results not only accurate, but also increasingly efficient in respect to various phenomena of scientific interest.

Among the range of options of numerical methods usually employed, some are particularly noteworthy, which in particular have a massive widespread use in solving ordinary and partial differential equations. The best known and most popular are: Finite Difference Method (FDM), Finite Element Method (FFM) and Finite Volume Method (FVM).

Several authors, such as in Refs. [107, 127], define the FVM as the integration, in space and time, of the differential equation in its conservative form in a control volume.

In this method the domain must be discretized into a finite number of control volumes, which must be continuous. Thus, the conservation equations are applied in each of the volumes that will be used in the integration process, as represented in Fig. 2.4. Thus, Eq. (2.2) can be rewritten as:

$$\begin{aligned}&\int_{t}^{t+\Delta t}\int_{s}^{n}\int_{w}^{e}\frac{\partial C}{\partial t}dxdydt+\int_{t}^{t+\Delta t}\int_{s}^{n}\int_{w}^{e}u\frac{\partial C}{\partial x}dxdydt=\\&\int_{t}^{t+\Delta t}\int_{s}^{n}\int_{w}^{e}D_l\frac{\partial}{\partial x}\left(\frac{\partial C}{\partial x}\right)dxdydt+\int_{t}^{t+\Delta t}\int_{s}^{n}\int_{w}^{e}D_t\frac{\partial}{\partial y}\left(\frac{\partial C}{\partial y}\right)dxdydt\end{aligned} \tag{2.11}$$

The time derivative term is expressed as

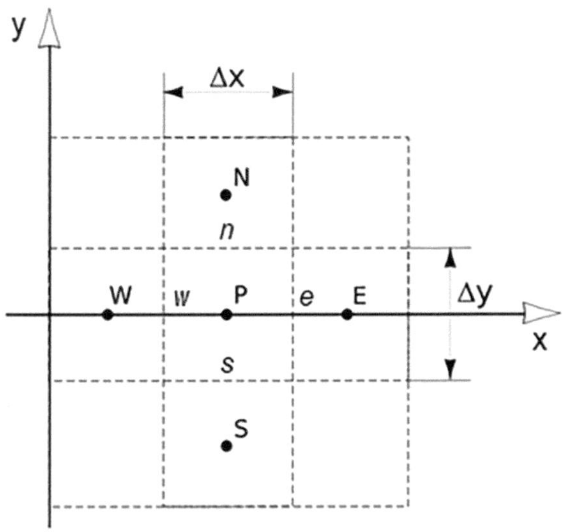

Fig. 2.4 Domain discretization for the application of the Finite Volume Method

$$\int_t^{t+\Delta t}\int_s^n\int_w^e \frac{\partial C}{\partial t}dxdydt = \Delta x\Delta y\left(C_P^{t+\Delta t} - C_P^t\right) \tag{2.12}$$

where Δx and Δy represent the space steps in the discretized domain, in the x and y coordinates, respectively.

Several interpolation schemes can be adopted for the approximation of derivatives of the advective term. In this chapter, we opted for the use of the upwind scheme, in which the value of the function in the interface is replaced by the values of the term upstream of the control volume. The nodes used in such scheme change according to the direction of the flow velocity.

Integrating the advective term of Eq. (2.11) in space yields:

$$\int_t^{t+\Delta t}\int_s^n\int_w^e u\frac{\partial C}{\partial x}dxdydt = \int_t^{t+\Delta t} u(C_e - C_w)\Delta y dt \tag{2.13}$$

Applying the upwind scheme in Eq. (2.13) we find:

$$\int_t^{t+\Delta t} u(C_e - C_w)\Delta y dt = \int_t^{t+\Delta t} u(C_E - C_P)\Delta y dt \tag{2.14}$$

The behavior of the concentration C on the faces of the elemental volume can be evaluated using the concentration at time t or $t + \Delta t$. In this way, an interpolation function describes the behavior of the concentration during the time interval Δt. In Ref. [191] this function is described as

$$C_P = \theta C_P^{t+\Delta t} + (1-\theta)C_P^t \tag{2.15}$$

where $0 \leq \theta \leq 1$ is a coefficient related to explicit, $\theta = 0$, and implicit, $\theta = 1$, formulations.

Integrating the longitudinal and transversal dispersion terms, we have:

$$\int_t^{t+\Delta t}\int_s^n\int_w^e D_l\frac{\partial}{\partial x}\left(\frac{\partial C}{\partial x}\right)dxdydt = \int_t^{t+\Delta t} D_l\left(\frac{\partial C}{\partial x}\Big|_e - \frac{\partial C}{\partial x}\Big|_w\right)\Delta y dt \tag{2.16}$$

$$\int_t^{t+\Delta t}\int_s^n\int_w^e D_t\frac{\partial}{\partial y}\left(\frac{\partial C}{\partial y}\right)dxdydt = \int_t^{t+\Delta t} D_t\left(\frac{\partial C}{\partial y}\Big|_n - \frac{\partial C}{\partial y}\Big|_s\right)\Delta x dt \tag{2.17}$$

Approximating the derivatives in Eqs. (2.16) and (2.17) with central differences, and using the interpolation function described by Eq. (2.15) for integration in time results

$$\begin{aligned}&\int_t^{t+\Delta t} D_l\left(\frac{C_E - 2C_P + C_W}{\Delta x}\right)\Delta y dt = \\ &D_l\left[\theta\left(\frac{C_E^{t+\Delta t} - 2C_P^{t+\Delta t} + C_W^{t+\Delta t}}{\Delta x}\right) + (1-\theta)\left(\frac{C_E^t - 2C_P^t + C_W^t}{\Delta x}\right)\right]\Delta y\Delta t\end{aligned} \tag{2.18}$$

$$\int_{t}^{t+\Delta t} D_t\left(\frac{C_N - 2C_P + C_S}{\Delta y}\right)\Delta x dt = \\ D_t\left[\theta\left(\frac{C_N^{t+\Delta t} - 2C_P^{t+\Delta t} + C_S^{t+\Delta t}}{\Delta y}\right) + (1-\theta)\left(\frac{C_N^t - 2C_P^t + C_S^t}{\Delta y}\right)\right]\Delta x\Delta t \tag{2.19}$$

where Δt represent the time step in the discretized domain.

The adopted values for the parameter θ indicate the type of approximation in the time that will be used in the equations and, consequently, the formulation to be used in the FVM. For the problem proposed in the present chapter, the totally implicit formulation ($\theta = 1$) is used because it is unconditionally stable, and of easy computational implementation. Thus, the following formulation is used for the FVM when the flow is in the x positive direction:

$$-D_t\left(\frac{\Delta x}{\Delta y}\right)C_N^{t+\Delta t} - D_l\left(\frac{\Delta y}{\Delta x}\right)C_W^{t+\Delta t} + \\ \left(\frac{\Delta x\Delta y}{\Delta t} + u\Delta y + 2D_l\frac{\Delta y}{\Delta x} + 2D_t\frac{\Delta x}{\Delta y}\right)C_P^{t+\Delta t} - \left(u\Delta y + D_l\frac{\Delta y}{\Delta x}\right)C_E^{t+\Delta t} + \\ -D_t\left(\frac{\Delta x}{\Delta y}\right)C_S^{t+\Delta t} = \frac{\Delta x\Delta y}{\Delta t}C_P^t \tag{2.20}$$

For the control volumes located at the boundaries, particular equations must be defined.

To simulate the instantaneous release of a pollutant or a tracer, e.g. a salt solution, a control volume was defined as the "release volume", in which the total mass of added salt was immediately homogenized. The resulting dimensions and, therefore, the concentration in this release volume varied according to the mesh adopted.

The discretization procedure results in a linear system of equations that can be expressed by a quite sparse matrix. Thus, taking Eq. (2.20) and imposing both boundary and initial conditions yield a system of linear algebraic equations, of the type $AC = b$.

The generated matrix can be solved through direct or iterative methods. For iterative methods, such as Gauss–Seidel, convergence is achieved when the Scarborough criterion is satisfied [191]

$$\frac{\sum |a_{nb}|}{|a_P|} \quad \begin{cases} \leq 1 \textit{ for all equations and} \\ < 1 \textit{ for at least one of the equations} \end{cases} \tag{2.21}$$

where a_P is the coefficient of the central volume P, and a_{nb} are coefficients of the neighboring volumes of the volume P.

The system $AC = b$ is solved through the Thomas algorithm as described in Refs. [107, 127, 171, 191].

2.4 Model Verification and Calibration

2.4.1 Mesh Variation

The numerical solution of the differential equations has truncation errors generated by the approximations of the derivatives using Taylor series [107, 157, 175].

As a result of these errors, the numerical solutions can present oscillations or false numerical diffusion, not satisfactorily representing the real physical phenomena. In the case discussed in this chapter, the use of an upwind approximation for the advective term may lead to false numerical diffusion.

Techniques for the minimization of false diffusion are presented in Refs. [56, 175], which take into account the Courant number, which is defined by:

$$\nu_{Co} = \frac{u\Delta t}{\Delta x} \tag{2.22}$$

Based on this, some simulations were performed in order to verify the best mesh to be used in the simulations presented in this chapter. In a first estimate, the behavior of a conservative substance in the Macaé River was evaluated considering the instantaneous release of 20 kg of common salt (NaCl) as a tracer, close to the right margin of the river, as represented by the solid square in Fig. 2.3, upstream the monitoring station, represented by the solid circle in the same figure. The mean width and depth in that river stretch were 42 m and 1 m, respectively. The transversal and longitudinal dispersion coefficients were calculated using Eqs. (2.7) and (2.9), respectively.

If the gravity acceleration is assumed to be 9.8 m s^{-2}, the mean depth as 1 m and river bed slope as 0.0005 m m^{-1}, results in the friction velocity described by Eq. (2.8) equal to 0.07 m s^{-1}, yielding, the following values for the transversal and longitudinal dispersion coefficients: $D_t = 0.01\ m^2\ s^{-1}$ and $D_l = 0.42\ m^2\ s^{-1}$.

Figure 2.5 shows the simulated salt concentration behavior adopting the parameters values listed in Table 2.1, considering different space steps, Δx and Δy. For $\Delta x = \Delta y =$ 2.8 m, 1.4 m, and 0.7 m, the domain was divided, respectively, into 975, 3,900 and 15,600 control volumes. It is verified that only the mesh with $\Delta x = \Delta y = 0.7$ m presents the Courant number equal to 1, i.e. $\nu_{Co} = 1$. It is observed a higher attenuation of the peak concentration when adopting this mesh.

2.4.2 Effect of the Transversal Dispersion

The effects arising from the variation of the transversal dispersive term are considered in this section. Here it is assumed that the transversal component of velocity is negligible, i.e. $v = 0$, in a way that the transversal transport is solely driven by dispersive mechanisms. The domain of calculation is 182 m long and 42 m wide, and is simulated an instantaneous release of a conservative substance at a given point

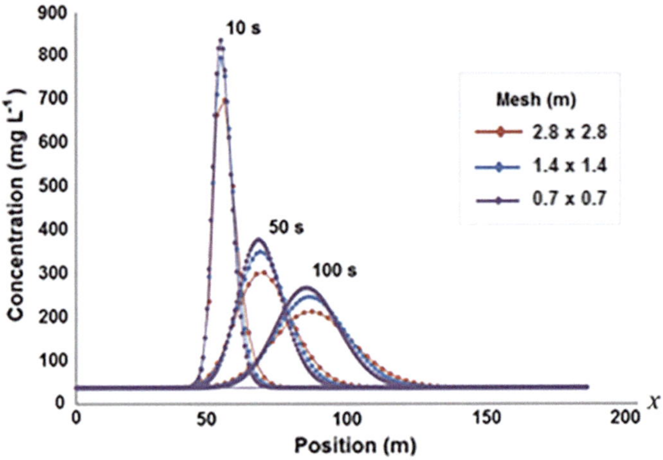

Fig. 2.5 Graph of the behavior of the concentration over time along the river bank for different meshes, at the position $y = 0.7$m. See Fig. 2.3

Table 2.1 Values of the parameters used in the simulations

Parameters	Value
Velocity, u [m s^{-1}]	0.35
Transversal dispersion coefficient, D_t[m^2 s^{-1}]	0.01
Longitudinal dispersion coefficient, D_l [m^2 s^{-1}]	0.42
L_x and L_y [m]	182.00 and 42.00
Launch position, x and y [m]	50.00 and 0.70
Time step, Δt [s]	2.00
Space step, $\Delta \boldsymbol{x} = \Delta \boldsymbol{y}$ [m]	0.70, 1.4 or 2.8

close to the river margin. The values of the longitudinal dispersion coefficient and the other information used in the simulations are described in Table 2.1, except for the transversal dispersion coefficient D_t and velocity u.

Figure 2.6 shows the simulated spatial distribution of the concentration at different instants of time, adopting a range of values for the transversal dispersion coefficient. In these simulations it was considered $u = 0$, the launch position at $x = 91.00$ m and $y = 0.70$ m. For any given time instant it is possible to observe a considerable decrease in the concentration of salt in response to the increase in the transversal dispersion coefficient.

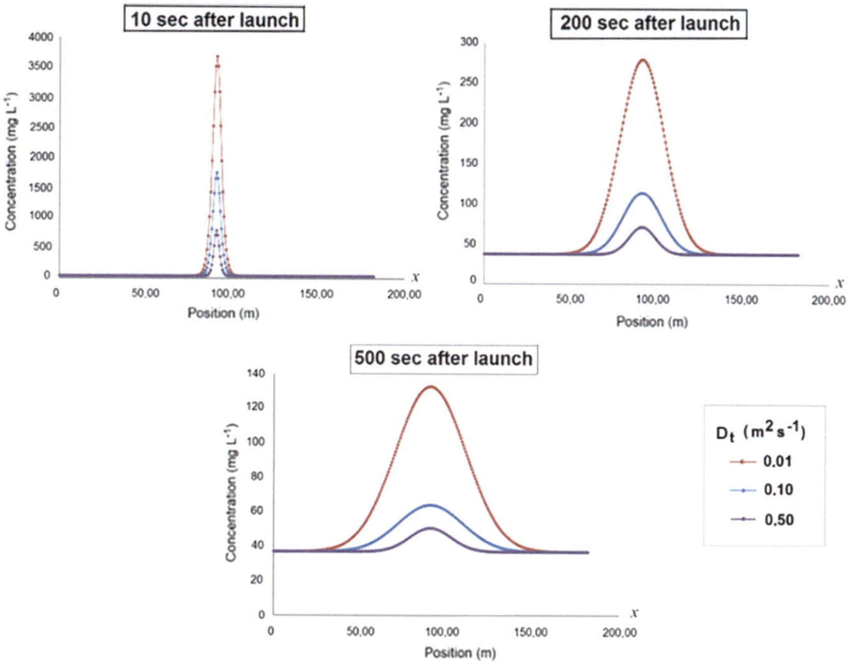

Fig. 2.6 Graph of the behavior of the concentration considering only the variation of the transversal dispersive effect at 0.70 m of the margin, with $u = 0$

2.4.3 *Effect of the Longitudinal Dispersion*

A similar evaluation was performed regarding the longitudinal dispersion coefficient. The longitudinal component of velocity was set to zero, i.e. $u = 0$, and the launch position at $x = 91.00$ m and $y = 0.70$ m. The values for the other parameters are the same as those used in the previous subsection, shown in Table 2.1, except for the longitudinal dispersion coefficient, D_l, and velocity u.

Figure 2.7 shows the simulated spatial distribution of the concentration at different instants of time, adopting a range of values for the longitudinal dispersion coefficient. As expected, the salt concentration dispersion becomes more effective as the longitudinal dispersion coefficient increases.

2.4.4 *Calibration*

As in the simulations in Sects. 2.4.2 and 2.4.3, in which there was only the variation of the transversal and longitudinal dispersion coefficients, respectively, the tracer launch occurred for the real experiment considered here at $x = 91$ m and at 0.70 m

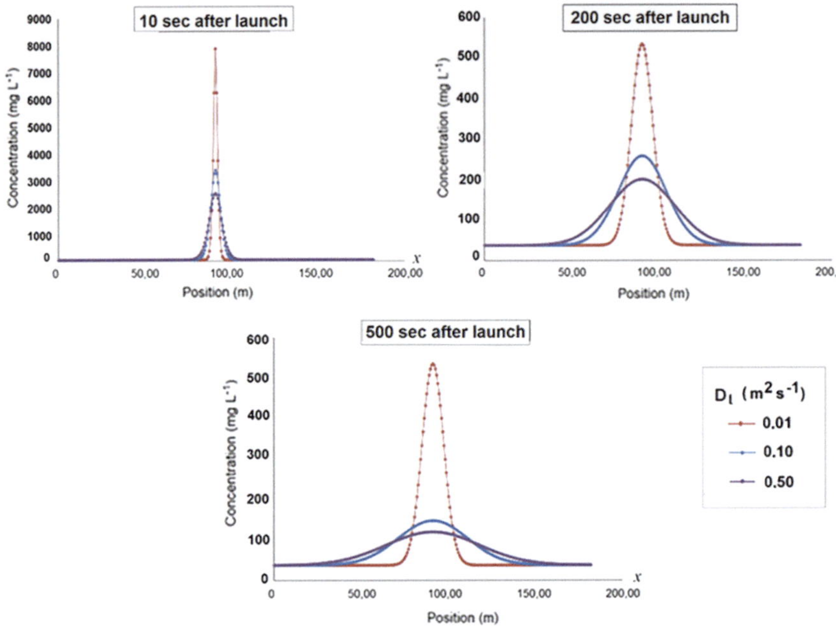

Fig. 2.7 Graph of the behavior of the concentration considering only the variation of the longitudinal dispersive effect at 0.70 m of the margin, with $u = 0$

from the margin with a concentration of 40.816 mg L^{-1} in the launch volume of the discretized computational domain.

The calibration of the model was performed by comparing the results of the simulations with those obtained in the field works. In this sense, numerical tests were carried out to determine the parameters that would lead to the best fit between the numerical solution and the experimental observations.

In a first moment, tests were performed varying the computational mesh used in the simulations. A comparison was made of the numerical results obtained with the meshes presented in Sect. 2.4, with those results obtained in the field work. The values adopted for the parameters in the simulations are described in Table 2.2.

The results of the plume concentration behavior at the collection point, for each mesh variation, are shown in Fig. 2.8.

As previously discussed, depending on the mesh adopted, the numerical solution may not accurately represent the physical phenomena studied due to the influence of truncation errors and false diffusion. These problems can be minimized by reducing the truncation errors and adopting a Courant number of 1. Thus, we chose to use the mesh with $\Delta x = \Delta y = 0.7$ m.

In order to obtain the best fit between the numerical solution and the observed data, variations in the longitudinal and transversal dispersion coefficients were performed. For the other parameters, the values are those shown in Table 2.2. The best fit of the numerical results in relation to the experimental data is reached for $D_t = 0.008$

Table 2.2 Values of the parameters used at each mesh variation simulation

Parameters	Value
Velocity, u [m s^{-1}]	0.35
Transversal dispersion coefficient, D_t [m^2 s^{-1}]	0.01
Longitudinal dispersion coefficient, D_l [m^2 s^{-1}]	0.42
L_x and L_y [m]	182.00 and 42.00
Launch position, x and y [m]	91.00 and 0.70
Collecting position, x and y [m]	100.00 and 0.70
Time step, Δt [s]	2.00
Simulation time, t_f [s]	352.00

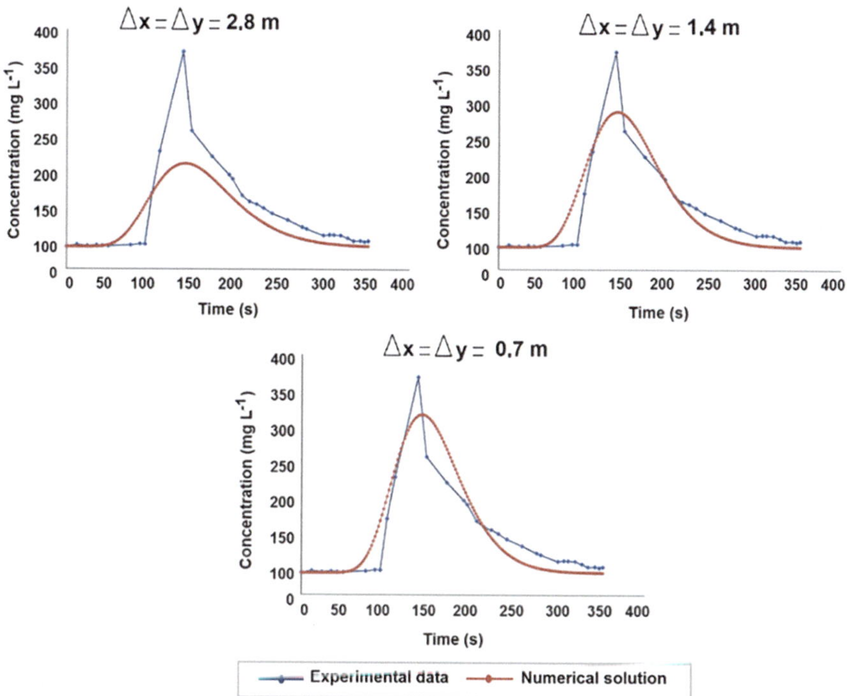

Fig. 2.8 Graph of the behavior of the concentration at the point of collection for each mesh space step considered

m^2 s^{-1} and $D_l = 0.33$ m^2 s^{-1} (Fig. 2.9). These values for the dispersion coefficients are well within the margin of error of the correlations given by Eqs. (2.7) and (2.9). A two-dimensional view of the plume at the moment its peak concentration reaches the sampling point, i.e. at $t = 136$ s, is shown in Fig. 2.10.

Adopting a two-dimensional simulation of contaminants advection and dispersion, i.e. vertically integrated, it may be interesting to establish the distance within

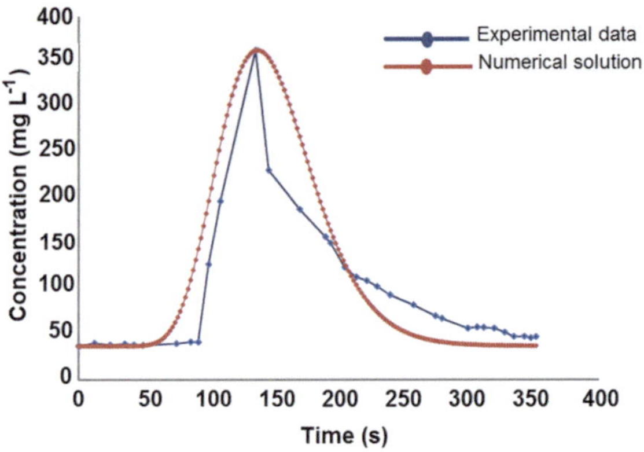

Fig. 2.9 Graph of the behavior of the concentration at the point of collection considering $\Delta x = \Delta y = 0.7$ m; $D_t = 0.008$ m^2 s^{-1} and $D_l = 0.33$ m^2 s^{-1}

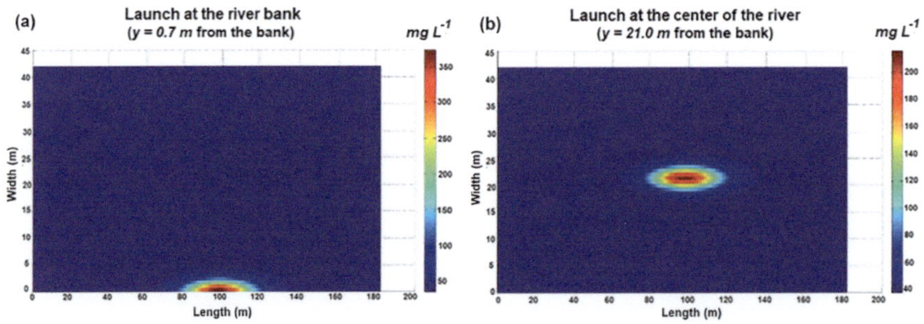

Fig. 2.10 Tracer behavior after 136 s from its launch at river bank (**a**) and at 21 m from the river bank (center of the river) (**b**), considering u = 0.35 m s^{-1}; $D_l = 0.33$ m^2 s^{-1} and $D_t = 0.008$ m^2 s^{-1}

which the transversal mixture in the transversal direction occurs [103]. From this point on, a one-dimensional approach (integrated in the cross-section) is enough to simulate this process, which may be convenient, from the computational cost point of view for instance. This distance known as the zone of mixing, or advective zone, is characterized by the domain of the advective effect. After this advective interval of non-homogenized mixing, the effects of advection are counterbalanced by transverse turbulent diffusion, and the concentration profile evolves to a Gaussian form. This effect is somewhat observed in Fig. 2.5.

According to Ref. [12], the length of this advective zone can be estimated through an equation adapted from Ref. [197],

$$L_{min} = K^* \frac{uB^2}{D_t} \tag{2.23}$$

where L_{min} is the length required for mixing, i.e. the mixing length, B is the average river width section and K^* the dimensionless coefficient whose value takes into account the degree of mixing, as well as the number and position of the injection points. The other mathematical symbols in Eq. (2.23) have already been defined.

For launches in the river axis and with a mixture around 95%, the value of K^* is estimated at 0.1 [53]. Thus, considering the width of the Macaé River in the simulated section as 42 m, velocity equal to 0.35 m s^{-1} and the transversal dispersion coefficient as 0.008 m^2 s^{-1}, the length required for the complete homogenization of the tracer mixture will be approximately 7.7 km. From this computation, it is obvious that the computational domain considered is insufficient for this complete mixture to occur if the discharge is located at $y = 21$ m, that is, in the center of the river.

For purposes of cost savings, i.e. reduction in the computational time, we opted here for a hypothetical launch using a coarser mesh where $\Delta x = \Delta y = 3.5$ m, and $\Delta t = 10$ s. Therefore, the Courant number remains equal to 1. The values of the parameters used in the simulation are shown in Table 2.3.

According to Ref. [109], if the discharge is located at $y = 0.7$ m, that is for riverbank launches—as addressed here—the length required for total homogenization of the concentration, i.e. the mixing length, is proportional to the square of twice the width of the river:

$$L_{\min} = 0.1\frac{u(2B)^2}{D_t} \tag{2.24}$$

and, therefore,

$$L_{\min} = 0.4\frac{uB^2}{D_t} \tag{2.25}$$

Considering a riverbank launch, it is found that the best value for the parameter K^*, in Eq. (2.23) would be 0.4, one can observe that the same value is found in Ref. [109], which would imply a length of the advective zone necessary for the complete homogenization of the tracer around 30.08 km, a value that significantly exceeds the meeting point of the Macaé River with the Atlantic Ocean.

Figure 2.11 shows the behavior of the tracer plume, after 10,000 s from its launch considering the coarser mesh configuration. It is found that the mixture

Table 2.3 Values of the parameters using a coarser mesh simulation

Parameters	Value
Velocity, u [m s^{-1}]	0.35
Transversal dispersion coefficient, D_t [m^2 s^{-1}]	0.008
Longitudinal dispersion coefficient, D_l [m^2 s^{-1}]	0.33
L_x and L_y [m]	4,550.00 and 42.00
Launch position, x [m]	91.00
Time step, Δt [s]	10.00
Space step, $\Delta x = \Delta y$ [m]	3.50

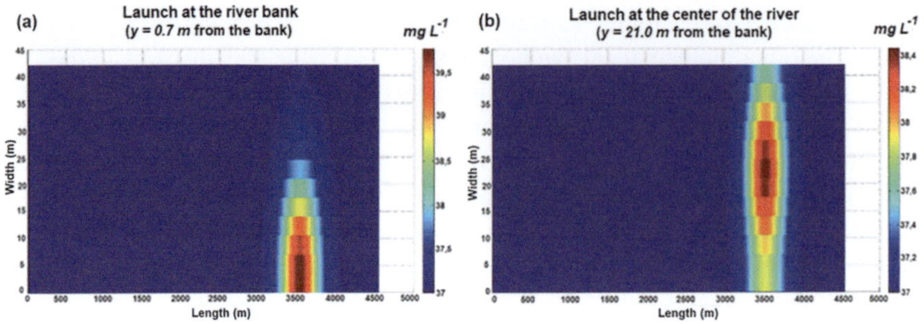

Fig. 2.11 Tracer behavior after 10,000 s from its launch at river bank (**a**) and at 21 m from river bank (center of river) (**b**), considering thicker mesh simulation

begins to homogenize in the transverse direction, with the dispersive effects having considerably attenuated the concentration peak.

It can be seen from Fig. 2.11a that the plume of the tracer over time is more dispersive in relation to the transversal direction, but the launch point is located very close to the river margin. As a result, the transversal direction, peak concentration is more pronounced when launching occurs at the river's banks rather than when it occurs in the center of the river (b). This behavior is expected because the mixing length for complete homogenization of the tracer is significantly higher for the former.

2.5 Modeling and Solving the Problem Using Artificial Intelligence

2.5.1 Modeling the Problem Using Artificial Neural Networks

An Artificial Neural Network (ANN) is a method of Artificial Intelligence composed of artificial neurons—which simulate human biological neurons—densely connected and exhibiting two distinct phases of operation: learning (training) and generalization.

The artificial neuron is a logical-mathematical structure that aims to simulate the shape, behavior, and functionality of a biological neuron. In this way, dendrites are replaced by inputs whose connections to the artificial cell body are made through elements called weights, which simulate synapses. The stimuli captured by the inputs are processed by the summation function, and the firing threshold of the biological neuron is represented by the transfer function [186]. Most neural network models have some training rule, where the weights and their connections are adjusted according to the presented patterns; they learn through examples [195].

Among the different neural topologies, it is considered that multilayer perceptron (MLP) networks, which use error backpropagation in their training, are the most frequently used due to their ease of implementation and less robust syntax.

To model the behavior of the concentration in the region of interest, aiming to simulate, by replacing the vertically integrated 2D transport model, the data obtained experimentally, the network was trained using the backpropagation algorithm and the Levenberg–Marquardt method for determining the weights and biases [182].

Due to the randomness in choosing the initial estimates of weights and biases, each simulation carried out in this work was processed around 30 times to find the best numerical results, using the experimental data obtained in the fieldwork as a reference.

In all the tests carried out, 33 pieces of information were passed as training patterns related to 6 parameters of significant influence on the modeling of the contaminant plume. These are: longitudinal launch position, transverse launch position, amount of tracer released into the river, longitudinal collection position, transverse collection position, and collection time.

With this information, the network can be characterized by a $P_{6\times 33}$ matrix as an input pattern, which maps a $T_{1\times 33}$ vector representing the concentrations from the fieldwork.

For the purpose of validating the training, the network was tested with a set of data that were not part of the training.

2.5.2 *Results of the Problem Using Artificial Neural Networks*

Comparisons were made between the experimental results and the simulations obtained through the 2D mathematical model solved numerically using the Finite Volume Method and the training of a neural network with mean squared error minimization by Levenberg–Marquardt.

For the simulations involving the mathematical model, the best configuration of parameters obtained in the calibration step (Sect. 2.4.4) was used, as shown in Table 2.4.

Table 2.4 Parameter values used in the 2D mathematical model

Parameters	Value
Velocity, u [m s^{-1}]	0.35
Transversal dispersion coefficient, D_t [m^2 s^{-1}]	0.008
Longitudinal dispersion coefficient, D_l [m^2 s^{-1}]	0.33
L_x and L_y [m]	182.00 and 42.00
Launch position, x and y [m]	91.00 and 0.70
Collecting position, x and y [m]	100.00 and 0.70
Time step, Δt [s]	2.00
Simulation time, t_f [s]	352.00

Regarding the Artificial Neural Network, simulations were conducted by varying the number of hidden layers (from one to five layers) as well as the number of neurons (from 5 to 50 neurons) in each layer, to obtain the best fit between the numerical results and the experimental data.

For the training of the network, the launch and collection positions in the longitudinal and transverse directions were defined as indicated in Table 2.4, while the initial concentration, launched instantaneously into the river, was 175 g/L. The time points at which the samples were collected were the same as those obtained in the fieldwork, whose concentration profile is shown in Fig. 2.9. Regarding validation, the network's behavior was checked every 60 s within the interval [0, 360].

Initially, the neural network was trained using only one hidden layer and varying the number of neurons in this layer. Table 2.5 shows the parameters, while Fig. 2.12 displays the results obtained with the best configuration for training the network with one hidden layer. It is observed that training the neural network with only one hidden layer did not satisfactorily model the concentration peak, obtaining a root mean square error (RMSE) of approximately 22 mg/L.

A new training of the neural network was conducted using two hidden layers and varying the number of neurons in each to obtain the best set of parameters that numerically represented the data obtained in the fieldwork. The validation was performed with the same set of data used for testing the network with only one hidden

Table 2.5 Configuration of the neural network with one hidden layer used in the simulations

Parameters	Value
Number of neurons in the hidden layer	30
Activation functions	Tangent/hyperbolic/linear
Method	Levenberg–Marquardt
Learning rate	0.001
Maximum number of epochs	5,000

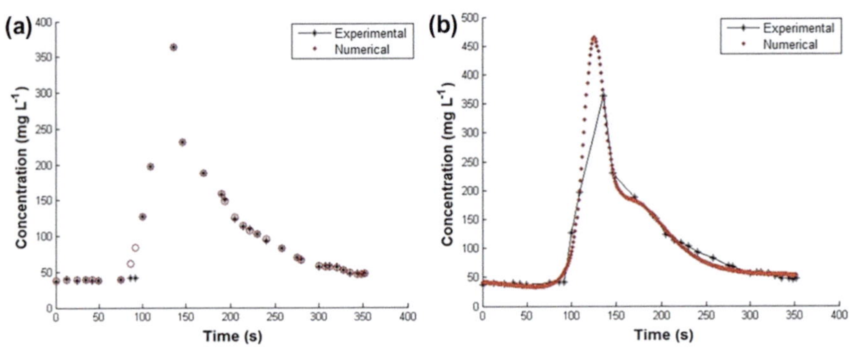

Fig. 2.12 Concentrations obtained at the collection point after training and testing the neural network with 30 neurons in the hidden layer, after 531 epochs. (**a**) Training. (**b**) Testing

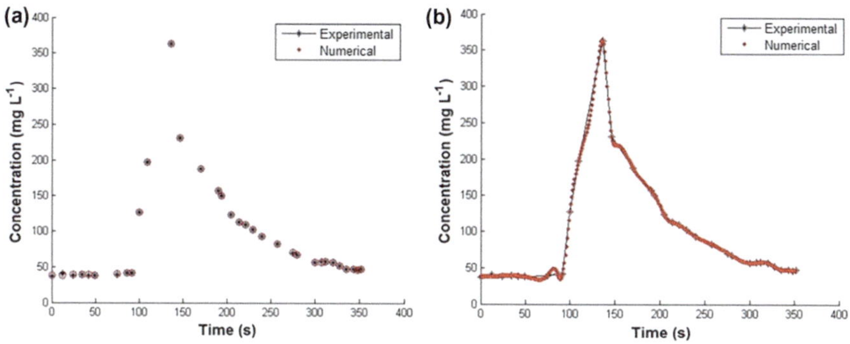

Fig. 2.13 Concentrations obtained at the collection point after training and testing the neural network with 14 and 8 neurons in the first and second hidden layers, respectively, after 1280 epochs. (**a**) Training. (**b**) Testing

layer. Figure 2.13 shows the results obtained with the training of the network with the best configuration, containing 14 and 8 neurons in the first and second hidden layers, respectively.

It can be observed in Fig. 2.13 that there is a better fit of the numerical result to the experimental data in terms of the concentration peak of the contaminant plume (tracer), having reached a root mean square error of around 1.7 mg/L.

2.6 Remarks

Several simulations were carried out, generating hypothetical scenarios for the injection of pollutant or tracer loads into rivers, allowing the evaluation of a real situation, i.e. an instantaneous launch in the low Macaé river.

According to the simulation results shown in Fig. 2.6, it was verified the response of the tracer plume behavior to the values attributed to the transversal dispersion coefficient and, as expected, it was found that the higher the coefficient, the greater the plume attenuation. Also, Fig. 2.7 allowed to evaluate the behavior of the concentration over time only in relation to the variation of the longitudinal dispersion coefficient. Analyzing the behavior of the plume, it is observed a considerable sensitivity with the variation of the coefficient of longitudinal dispersion, which increases with time. There was also a great sensitivity of the amplitude of the concentration curve to small variations of the dispersion coefficients, both transversal and longitudinal.

Simulations were carried out adopting the configuration of different computational meshes in order to minimize the effects of the truncation error, as well as a variation of the dispersive and advective terms, in order to verify the best numerical fit to the experimental data collected in the field work.

In view of the simulations, a trial-and-error approach was used in the model calibration, and the parameters that better adjusted the computational model were the ones that generated the results presented in Fig. 2.9, showing that the values observed and calculated numerically shows a larger discrepancy after the peak concentration is achieved. This may be due to a total non-homogenization of the concentration in the vertical direction or some physical decreasing effect, such as sedimentation, that was not contemplated in the simulation, which motivates future investigations.

In order to verify the best point of launching of tracers in streams, a comparison was made between an instantaneous injection of the tracer in the center of the river and another one close to its margin. Thus, by analyzing Fig. 2.10, it was found that a lower concentration peak occurs for the central launch compared to the experiment considering the margin launch, in addition to presenting a more pronounced dispersive response. It is concluded that a launch in this central position is less harmful to the environment compared to the same launch at the margin, which is confirmed in Fig. 2.11.

Regarding Artificial Neural Networks as a substitute for a 2D mathematical model of pollutant transport, it is observed that the training of the network with one and two hidden layers presented a good correlation with the values used for this purpose. However, when using data that were not part of the training, it is concluded that the network with only one hidden layer did not satisfactorily model the concentration peak of the contaminant plume, as shown in Fig. 2.12, unlike the training using two hidden layers, which achieved a better numerical fit in relation to the referred data, as shown in Fig. 2.13. This shows that the network trained with only one hidden layer was not able to generalize the modeling of the contaminant plume when using a set of patterns that were not part of its training.

Finally, it should be noted that the formulation and modeling used, despite adopting some simplifications, were able to satisfactorily represent the pollutant dispersion phenomena, modeling the concentration peak with acceptable accuracy. As a result, the approach adopted here presents itself as a valuable tool in the process of evaluating the environmental impact of effluent discharges into water bodies.

Chapter 3
Estimation of Pollutants Release Time Dependence in Fluvial Environments

The present chapter analyzes the transport and dispersion of a conservative substance in a river. A hybrid numerical-analytical methodology, based on integral transforms, is used to solve the one-dimensional advection–diffusion transient problem. The main focus of this chapter involves not only the evolution of effluent concentrations over time, but also the determination of the function that represents the time dependence of the launch of a contaminant in an aquatic environment, through the formulation of an inverse problem. The proposed model can be used as a tool for the analysis of environmental impacts due to the release of pollutants in a water body. Once the hydrodynamic parameters are known, it is possible to analyze the transport and the evolution of the pollutant plume, allowing the estimation of the areas that can be reached, avoiding thus, more serious damage to the environment. The direct problem is solved by the analytical–numerical hybrid method known as the Generalized Integral Transform Technique (GITT). In contrast, the inverse problem is formulated via Bayesian inference, and solved with the Markov Chain Monte Carlo Method (MCMC).

3.1 Introduction

Pollutants, when they reach water bodies, are influenced by several physical, chemical and biological mechanisms present in nature that can change their behavior and their concentrations [176, 188, 192].

Bruno Carlos Lugão, from the Secretary of Education of the State of Rio de Janeiro, and Luiz Alberto da Silva Abreu and Diego Campos Knupp, both from the Polytechnic Institute of the Rio de Janeiro State University, co-authored this chapter.

J. Lugon Junior et al., *Inverse Problems in the Environmental Modelling of Rivers and Estuaries*, Studies in Systems, Decision and Control 246,
https://doi.org/10.1007/978-3-032-09937-2_3

In order to facilitate the understanding of the phenomena of dispersion of a plume of contaminants in a water medium, some essential concepts related to the theme will be presented next.

Advection: is the transport of a substance that occurs as a function of the velocity field of a fluid [143, 188]. The displacement of a pollutant is directly related to the speed of the water body, the greater the velocity, the faster the contaminant plume moves away from the launch site.

Constituent: Eiger [46] defines constituent as a quantity, which, in the study of the dispersion of contaminants, characterizes a pollutant or a tracer. It can be measured by its concentration and is classified as:

- Conservative: the spatial and temporal variation of the constituent is not influenced by reactions with other constituents or with the fluid that contains it. Its distribution varies only by the physical processes of transport.
- Non-conservative: the spatial and temporal variation of the constituent is altered by reactions with other constituents or with the fluid that contains it.
- Active: constituent that affects the hydrodynamic characteristics of the flow.
- Passive: does not change the hydrodynamic characteristics of the flow.

The concentration of pollutants, when released into rivers, in addition to being affected by the advection is also affected by the diffusive processes, namely [53]:

- Molecular diffusion: is the scattering of a solute in a fluid due to the random motion of the molecules. This process is governed by Fick's law.
- Turbulent diffusion: is the random scattering of the particles due to the turbulent movement of the flow. This process is similar to molecular diffusion, but occurs at a higher rate.
- Dispersion: is the joint action of molecular and/or turbulent diffusion along with advection. The substance, in addition to undergoing displacement due to the velocity field of the river, diffuses spatially.
- Complete Mixing Zone: is the distance required for a constituent to be completely mixed throughout the cross-section of the flow. In Fig. 3.1 it is possible to observe a schematic representation for this process. In position x_1 is simulated the release of a substance in a continuous medium. In x_2 we observe a greater mixture of the substance, but it is not yet completely homogenized. Finally, after a certain distance $L_{\min}$, the substance has a completely mixed concentration profile in relation to the cross-flow section.

3.2 Direct Problem Formulation and Solution

The direct problem used in this chapter is intended for the analysis of the behavior of a pollutant released in a river. This phenomenon can be modeled by the equation of one-dimensional advection of a transient regime, with the mean velocity of the flow, the longitudinal dispersion coefficient, the initial condition and the boundary conditions being known.

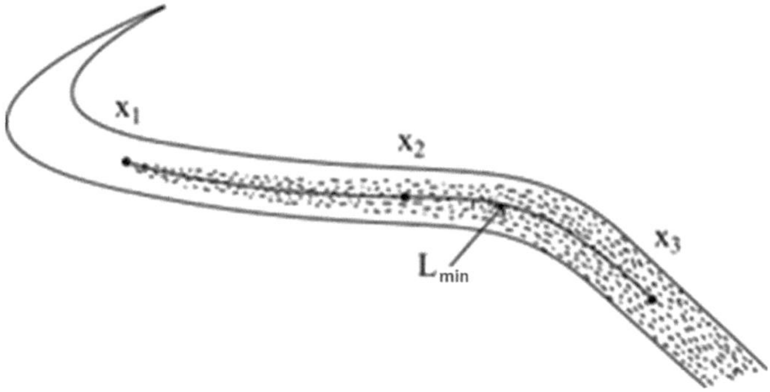

Fig. 3.1 Complete mixing zone. Adapted from [53]

According to Fick's first law, in the process of molecular diffusion, the mass flow of a substance is proportional to its concentration gradient, mathematically written as [17, 30, 143]:

$$J_d = -D\frac{\partial C}{\partial x} \tag{3.1}$$

where J_d is the diffusive mass flux; D is the molecular diffusion coefficient; C is the concentration and x is the direction in which the process is developed.

It is important to point out that the minus sign in Eq. (3.1) is intended to ensure the positive mass transport from the higher concentration region to the lower one. Observe that in such situation the concentration gradient $\frac{\partial C}{\partial x}$ is negative.

Considering Eq. (3.1) along with a mass balance in a control volume [17, 164], it is possible to find the relation that represents the way the mass behaves in the process of Fickian diffusion. This relation, Fick's second law, is known as the diffusion equation, and is mathematically written, for the one-dimensional case, as [14, 17]

$$\frac{\partial C}{\partial t} = D\frac{\partial^2 C}{\partial x^2} \tag{3.2}$$

In the literature there are many works focused on the development of analytical solutions for Eq. (3.2). A classical solution, which comprises a case of interest in environmental studies, occurs when a pollutant is released instantaneously, at $t = 0$, in a water body. This scenario can be modeled through the following initial and boundary conditions in an infinite medium [182]:

$$C(x, 0) = \frac{M}{A_c}\delta(x) \tag{3.3}$$

$$C(\pm\infty, t) = 0 \tag{3.4}$$

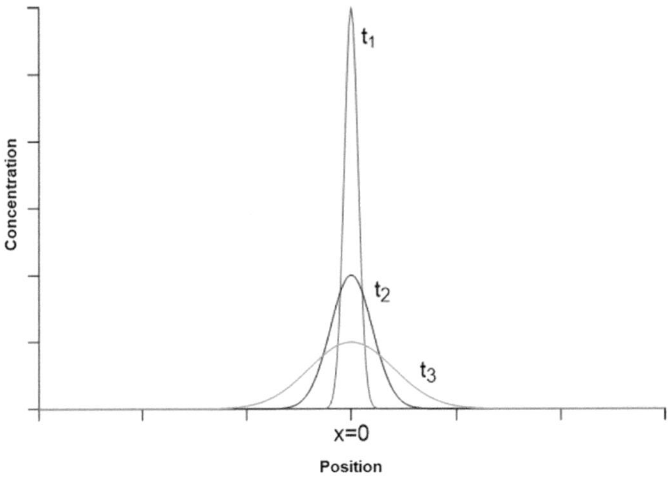

Fig. 3.2 Solution of the diffusion equation in an infinite medium, Eq. (3.5), for an instantaneous emission of pollutant

where M indicates the mass of the pollutant, A_c is the river cross-sectional area, and $\delta(x)$ is the Dirac delta function.

Chapra [30] presents the solution for Eqs. (3.2–3.4) as

$$C(x, t) = \frac{M}{A_c\sqrt{4\pi Dt}} \exp\left(-\frac{x^2}{4Dt}\right) \tag{3.5}$$

Figure 3.2 shows the behavior of a pollutant released instantaneously in a water medium according to Eq. (3.5) for three different time instants, with $t_1 < t_2 < t_3$. The concentration of the substance decreases according to a Gaussian profile, since, by the equation, Dt works as the variance of a normal distribution that increases over time.

Note that Eq. (3.2) takes into account only the diffusive processes. In order to insert the mass transport due to the advection phenomenon, that is, the macroscopic amount of mass transported through the fluid velocity field, the advective flux can be modeled by [14]:

$$J_a = u\,C \tag{3.6}$$

where J_a is the advective mass flux in the x direction, and u is the velocity of the flow in the x direction.

By performing an analysis for a control volume, and applying the principle of mass conservation, it is possible to describe the one-dimensional advection–diffusion equation in a transient regime as follows,

$$\frac{\partial C}{\partial t} + u\frac{\partial C}{\partial x} = D\frac{\partial^2 C}{\partial x^2} \tag{3.7}$$

Equation (3.7) can also be used to model the behavior of the instantaneous release of a substance into a water body, now under the influence of advection. Considering the initial and boundary conditions given by Eqs. (3.3) and (3.4), respectively, in Ref. [30] it is presented the following analytical solution,

$$C(x,t) = \frac{M}{A_c\sqrt{4\pi Dt}}\exp\left(-\frac{(x-ut)^2}{4Dt}\right) \tag{3.8}$$

Figure 3.3 represents the behavior of the solution of Eq. (3.7) for an instantaneous pollutant emission at $t = 0$, using Eq. (3.8), for three different positions, with $x_1 < x_2 < x_3$. It is possible to notice that, such as in the solution given by Eq. (3.5), the concentrations follow a varying Gaussian profile, as in Fig. 3.2, but, in this case, also with the solute transport resulting from the advective effect.

In order to understand the effects resulting from this type of transport, the instantaneous quantities are considered as the sum of their mean temporal value and a turbulent fluctuation resulting from the random oscillations of the particles of the fluid. Thus,

$$C = \overline{C} + C' \tag{3.9a}$$

$$u = \overline{u} + u' \tag{3.9b}$$

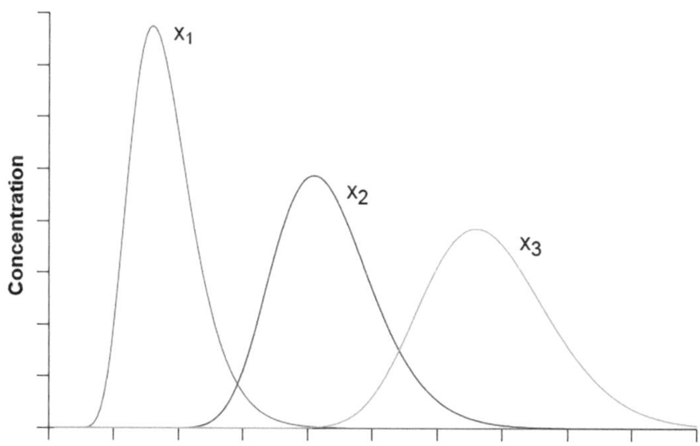

Fig. 3.3 Solution of the advection–diffusion equation in an infinite medium, Eq. (3.7), for an instantaneous emission of pollutant

with

$$\overline{C} = \frac{1}{\Delta t} \int_{t}^{t+\Delta t} C dt \tag{3.10a}$$

$$\overline{u} = \frac{1}{\Delta t} \int_{t}^{t+\Delta t} u dt \tag{3.10b}$$

where $\overline{C}$ and $\overline{u}$ represent the time averages of C and u, respectively, and C' and u' represent their turbulent fluctuations.

Applying Eqs. (3.9a, 3.9b) in Eq. (3.7), taking the time average in the resulting equation, and assuming the null averages of turbulent fluctuations and constant time averages in the integration interval results [175],

$$\frac{\partial \overline{C}}{\partial t} + \overline{u}\frac{\partial \overline{C}}{\partial x} = D\frac{\partial^2 \overline{C}}{\partial x^2} - \frac{\partial \overline{u'C'}}{\partial x} \tag{3.11}$$

Using an analogy with Fick's law, the term representing the turbulent advective transport in Eq. (3.11) can be transformed into a diffusive term as follows

$$\overline{u'C'} = -\varepsilon_x \frac{\partial \overline{C}}{\partial x} \rightarrow \frac{\partial \overline{u'C'}}{\partial x} = -\varepsilon_x \frac{\partial^2 \overline{C}}{\partial x^2} \tag{3.12}$$

where ε_x is the longitudinal turbulent diffusion coefficient. In this way, replacing Eq. (3.12) in Eq. (3.11), yields

$$\frac{\partial \overline{C}}{\partial t} + \overline{u}\frac{\partial \overline{C}}{\partial x} = D\frac{\partial^2 \overline{C}}{\partial x^2} + \varepsilon_x \frac{\partial^2 \overline{C}}{\partial x^2} = (D + \varepsilon_x)\frac{\partial^2 \overline{C}}{\partial x^2} \tag{3.13}$$

It is possible to observe the presence of terms representing molecular and turbulent diffusion in Eq. (3.13). According to Ref. [12], the influence of the term that represents the turbulent diffusion in the scattering of a substance in a hydric body is much more intense than that originating from the molecular diffusion ($\varepsilon_x \gg D$).

After the complete mixing of the substance occurs in the cross-section of the flow, the concentration change occurs mainly in the longitudinal direction. From that moment on, the concentration is more strongly altered by the transverse velocity gradients [12]. In order to insert the effects of these gradients, the time average values of velocity and concentration are written in terms of the mean values in the cross section as follows

$$\overline{C} = C_m + C'' \tag{3.14a}$$

$$\overline{u} = u_m + u'' \tag{3.14b}$$

with

$$C_m = \frac{1}{A_c}\int \overline{C} dA \tag{3.15a}$$

$$u_m = \frac{1}{A_c}\int \overline{u} dA \tag{3.15b}$$

where C_m and u_m represent the cross section mean value of $\overline{C}$ and $\overline{u}$, respectively, and C'' and u'' represent the spatial fluctuations of $\overline{C}$ and $\overline{u}$ in the cross section, with area A_c.

Replacing Eqs. (3.14a, 3.14b) in Eq. (3.13), and taking the spatial average in the resulting equation, results, as demonstrated in Ref. [146], that

$$\frac{\partial C_m}{\partial t} + u_m \frac{\partial C_m}{\partial x} = (D + \varepsilon_x)\frac{\partial^2 C_m}{\partial x^2} - \frac{\partial \overline{\overline{u''C''}}}{\partial x} = \frac{\partial}{\partial x}\left[(D + \varepsilon_x)\frac{\partial C_m}{\partial x} - \overline{\overline{u''C''}}\right] \tag{3.16}$$

In Eq. (3.16), the term in brackets, similarly to Fick's law, is assumed proportional to the longitudinal concentration gradient, that is,

$$(D + \varepsilon_x)\frac{\partial C_m}{\partial x} - \overline{\overline{u''C''}} = D_l \frac{\partial C_m}{\partial x} \tag{3.17}$$

being D_l the longitudinal dispersion coefficient, and the double bar indicates the spatial mean value. By replacing Eq. (3.17) in Eq. (3.16), we finally have the expression of the equation used in this chapter, known as the advection–dispersion equation, in its one-dimensional form

$$\frac{\partial C_m}{\partial t} + u_m \frac{\partial C_m}{\partial x} = D_l \frac{\partial^2 C_m}{\partial x^2} \tag{3.18a}$$

From this point on, in order to simplify the notation, C_m and u_m will be replaced by C and u, respectively, and therefore, Eq. (3.18a) is rewritten as

$$\frac{\partial C}{\partial t} + u \frac{\partial C}{\partial x} = D_l \frac{\partial^2 C}{\partial x^2} \tag{3.18b}$$

Furthermore, modelling biochemical processes in terms of first-order reactions, a transformation rate directly proportional to the solute concentration is included in Eq. (3.18b), leading to the following advection–dispersion-reaction equation [138]

$$\frac{\partial C}{\partial t} + u \frac{\partial C}{\partial x} = D_l \frac{\partial^2 C}{\partial x^2} - \lambda C \tag{3.18c}$$

where λ is the first-order rate coefficient. For conservative substances, the value of this parameter is assumed to be null.

As in Eqs. (3.2) and (3.7), Eq. (3.18b) can also be used to model the instantaneous release of a pollutant into a river, at $t = 0$, by considering the conditions given by Eqs. (3.3) and (3.4). The solution found for this case is identical to that represented by Eq. (3.8), but now the parameter representing molecular diffusion is replaced by the longitudinal dispersion coefficient.

Another scenario of interest in the study of the behavior of contaminants in rivers occurs when considering the continuous launch of a substance with concentration C_0 during a given time period $[0, \tau]$. This case can be represented by the following initial and boundary conditions in a semi-infinite domain, i.e. $x \geq 0$,

$$C(x, 0) = 0, \quad \text{for } 0 \leq x < \infty \tag{3.19}$$

$$C(0, t) = \{0 \quad \text{if } t > \tau, \quad C_0 \quad \text{if } 0 < t \leq \tau\} \tag{3.20}$$

$$C(\infty, t) = 0 \quad \text{for } t > 0 \tag{3.21}$$

In Ref. [138] it is presented a solution to Eq. (3.18c), divided in two parts, considering the conditions given by Eqs. (3.19–3.21). The first one is valid during the pollutant launch time, i.e., from $t = 0$ until $t = \tau$,

$$\begin{aligned} C(x, t) = \frac{C_0}{2}\Bigg\{&\exp\left[\frac{ux}{2D_l}(1 - \Gamma)\right] erfc\left(\frac{x - ut\Gamma}{2\sqrt{D_l t}}\right) \\ &+ \exp\left[\frac{ux}{2D_l}(1 + \Gamma)\right] erfc\left(\frac{x + ut\Gamma}{2\sqrt{D_l t}}\right)\Bigg\} \end{aligned} \tag{3.22}$$

where

$$\Gamma = \sqrt{1 + 2H} \tag{3.23a}$$

$$H = 2\lambda \frac{D_l}{u^2} \tag{3.23b}$$

being *erfc* the complementary error function.

The second part of the solution, used for $t > \tau$, is given by

$$\begin{aligned} C(x, t) = \frac{C_0}{2}\Bigg\{&\exp\left[\frac{ux}{2D_l}(1 - \Gamma)\right]\left[erfc\left(\frac{x - ut\Gamma}{2\sqrt{D_l t}}\right) - erfc\left(\frac{x - u(t - \tau)\Gamma}{2\sqrt{D_l(t - \tau)}}\right)\right] \\ &\times\left[erfc\left(\frac{x + ut\Gamma}{2\sqrt{D_l t}}\right) - erfc\left(\frac{x + u(t - \tau)\Gamma}{2\sqrt{D_l(t - \tau)}}\right)\right]\Bigg\} \end{aligned} \tag{3.24}$$

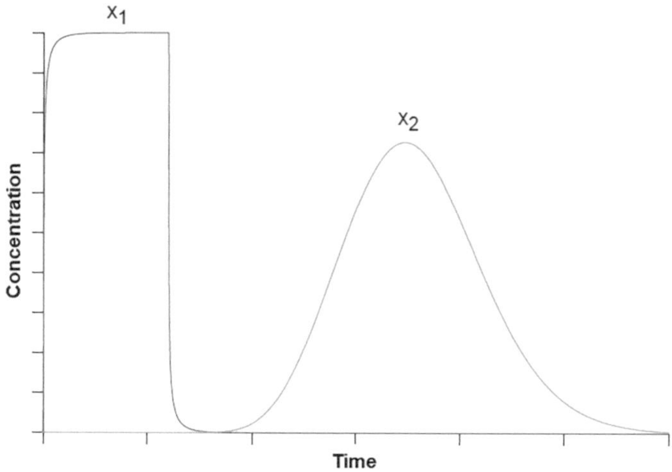

Fig. 3.4 Solution of the advection–dispersion equation considering the continuous pollutant emission of finite duration. For the location x_1 it is considered the case with $t < \tau$, Eq. (3.22), and for x_2, $t > \tau$, Eq. (3.24). It was considered also $\lambda = 0$ in Eq. (3.18c)

Figure 3.4 shows the behavior of the solution provided by Ref. [138] for the continuous launch of a substance of finite duration, i.e. during a time interval from 0 to τ. At x_1 which is located in the vicinity of $x = 0$, using Eq. (3.22) considering $t < \tau$, the concentration of the substance increases rapidly until it reaches the value of C_0. On the other hand, in a position x_2, at a greater distance from the boundary, using Eq. (3.24), considering $t > \tau$, the concentration of the substance follows more closely a Gaussian profile, as in Eqs. (3.5) and (3.8).

Equations (3.22) and (3.24) will be used for the validation of the solution of the direct problem obtained with the Generalized Integral Transform Technique (GITT). The GITT uses a basis of eigenfunctions obtained from an auxiliary eigenvalue problem and a transform-inverse pair. Once the eigenfunctions and their respective eigenvalues are obtained, the integral transform operator is applied to obtain an infinite system of coupled ordinary differential equations for the transformed potentials. This system is truncated in a number of terms which is considered sufficient for the required accuracy to be achieved, since it is usually possible to solve it only by means of numerical approaches. Given the transformed potentials and the basis of eigenfunctions, it is enough to apply the inverse transform to obtain the potential sought.

In the GITT formulation developed, a priori non-transformable terms are considered as a source of the equation used. Here we consider the problem of dispersion of pollutants in a water body, of length L, which is mathematically modeled by the advection–dispersion equation in transient regime, with the corresponding initial and boundary conditions:

$$\frac{\partial C(x,t)}{\partial t} + u\frac{\partial C(x,t)}{\partial x} = D_l\frac{\partial^2 C(x,t)}{\partial x^2}, \quad 0 < x < L, t > 0 \tag{3.25a}$$

$$C(0,t) = C_0(t), \quad t > 0 \tag{3.25b}$$

$$C(L,t) = 0, \quad t > 0 \tag{3.25c}$$

$$C(x,0) = C_1(x), \quad 0 \le x \le L \tag{3.25d}$$

The problem shown in Eqs. (3.25a–3.25d) is not homogeneous. According to Ref. [40] it is always interesting to reduce the importance of source terms and boundary conditions in order to improve the convergence of the expansions using GITT.

A possible alternative to overcome this difficulty is the use of analytical filters. They are expressions that can remove or reduce the degree of information not only from the boundary, but also from the source terms of a partial differential equation (PDE). According to Refs. [84, 85] several types of filters can be used in the same problem, relying on the user's experience in finding the best combination for each case.

Thus, before applying GITT, it will be necessary to find a filter that homogenizes the boundary condition given by Eq. (3.25b). Suppose the desired potential, $C(x,t)$, can be written as the sum of a filtered potential $C^*(x,t)$ with a filter $C_f(x,t)$, in other words:

$$C(x,t) = C^*(x,t) + C_f(x,t) \tag{3.26}$$

By performing the necessary derivatives in Eq. (3.26) and substituting in Eqs. (3.25a–3.25d), it is possible to formulate two distinct problems, one for the transformed potential and one for the filter. For the filter problem we have:

$$u\frac{\partial C_f(x,t)}{\partial x} - D_l\frac{\partial^2 C_f(x,t)}{\partial x^2} = 0, \quad 0 < x < L, \quad t > 0 \tag{3.27a}$$

$$C_f(0,t) = C_0(t), \quad t > 0 \tag{3.27b}$$

$$C_f(L,t) = 0, \quad t > 0 \tag{3.27c}$$

Note that the above differential equation has an almost steady-state formulation, in which the time t is considered as a parameter. With the given boundary conditions, the following solution is obtained for Eqs. (3.27a–3.27c):

$$C_f(x,t) = \frac{C_0(t)}{e^{\frac{Lu}{D_l}} - 1}\left(e^{\frac{Lu}{D_l}} - e^{\frac{xu}{D_l}}\right) \tag{3.28}$$

Once the filter expression is known, the filtered potential problem can be formulated so that it is homogeneous in the boundary as follows:

$$\frac{\partial C^*(x,t)}{\partial t} = D_l \frac{\partial^2 C^*(x,t)}{\partial x^2} + P(x,t), \quad 0 < x < L, \quad t > 0 \tag{3.29a}$$

$$C^*(0,t) = C_0(t) - C_f(0,t) = 0, \quad t > 0 \tag{3.29b}$$

$$C^*(L,t) = -C_f(L,t) = 0, \quad t > 0 \tag{3.29c}$$

$$C^*(x,0) = C_1(x) - C_f(x,0), \quad 0 \le x \le L \tag{3.29d}$$

$$P(x,t) = -u\frac{\partial C^*(x,t)}{\partial x} - \frac{\partial C_f(x,t)}{\partial t} \tag{3.29e}$$

It is important to note that the advective term and the filter time derivative present in Eq. (3.29e) are considered as a source term $P(x, t)$. It is now possible to apply the Generalized Integral Transform Technique to the filtered potential to improve convergence since its boundary conditions are homogeneous. When applying the GITT formalism, it is first necessary to define an integral transform pair in terms of an expansion on an eigenfunction basis as follows:

$$\textit{Integral Transform}: \quad \tilde{C}_i(t) = \int_0^L \tilde{\psi}_i(x) C^*(x,t)\,dx \tag{3.30}$$

$$\textit{Inverse Transform}: \quad C^*(x,t) = \sum_{i=1}^{\infty} \tilde{\psi}_i(x)\tilde{C}_i(t) \tag{3.31}$$

where $\tilde{C}_i(t)$ are the transformed potentials, and $\tilde{\psi}_i(x)$ are the normalized eigenfunctions, whose expression is given by:

$$\begin{aligned} \tilde{\psi}_i(x) &= \frac{\psi_i(x)}{\sqrt{N_i}} \\ N_i &= \int_0^L \psi_i^2(x)\,dx \end{aligned} \tag{3.32}$$

The eigenfunctions $\psi_i(x)$ are obtained from a Sturm–Liouville eigenvalue auxiliary problem [38], formulated from the application of the separation of variables method to the homogeneous problem represented by Eqs. (3.29a–3.29e), and solved only for the variable x, which results in:

$$\frac{d^2\psi_i(x)}{dx^2} + \frac{\mu_i^2}{D_l}\psi_i(x) = 0 \tag{3.33a}$$

$$\psi_i(0) = \psi_i(L) = 0 \tag{3.33b}$$

where μ_i are the eigenvalues and $\psi_i(x)$ their respective eigenfunctions. It is important to note that when considering a constant longitudinal dispersion coefficient D_l, the problem defined by Eqs. (3.33a, 3.33b) has an explicit analytical solution.

Using the operator $\int_0^L (.)\tilde{\psi}_i(x)dx$ in Eqs. (3.29a–3.29e), and performing algebraic manipulations on each resulting term, as can be seen in Ref. [38], yields the following system of ordinary differential equations (ODE) for the transformed potential $\tilde{C}_i(t)$:

$$\frac{d\tilde{C}_i(t)}{dt} + \mu_i^2\tilde{C}_i(t) = g_i(t), \quad i = 1, 2, 3, \ldots \tag{3.34a}$$

$$g_i(t) = -u\sum_{j=1}^{\infty}\tilde{C}_j(t)D_{ij} - G_i \tag{3.34b}$$

$$D_{ij} = \int_0^L \tilde{\psi}_i(x)\frac{d\tilde{\psi}_j(x)}{dx}dx \tag{3.34c}$$

$$G_i = \int_0^L \tilde{\psi}_i(x)\frac{\partial C_f(x,t)}{\partial t}dx \tag{3.34d}$$

$$\tilde{C}_i(0) = \int_0^L \tilde{\psi}_i(x)f^*(x)dx \tag{3.34e}$$

where $f^*(x) = \left[C_1(x) - C_f(x,0)\right]$.

The system of ordinary differential equations given by Eqs. (3.34a–3.34e) has no analytical solution, and, therefore, a numerical method is required. In practice, due to the impossibility of solving an infinite system, the transformed potentials are truncated in a certain order, N_{tr}, large enough for the desired accuracy to be achieved. In this work, the above system solution was obtained through the NDSolve routine of Mathematica software [193].

Once the transformed potentials, $\tilde{C}_i(t)$, and the normalized eigenfunctions, $\tilde{\psi}_i(x)$, are known, it is possible to use the inverse transform given by Eq. (3.31), to write the filtered potential $C^*(x,t)$. In this way, the solution of the problem defined with Eqs. (3.25a–3.25d) can be represented as:

$$C(x,t) = C^*(x,t) + C_f(x,t) = \sum_{i=1}^{N_{tr}}\tilde{\psi}_i(x)\tilde{C}_i(t) + C_f(x,t) \tag{3.35}$$

3.3 Inverse Problem Formulation

The physical problem studied consists of a conservative substance release in a hypothetical fluvial course, sufficiently shallow and narrow enough to justify a one-dimensional approach, which can be adequately formulated by the advection–dispersion equation, i.e. Eq. (3.18c) with $\lambda = 0$, which leads to Eq. (3.25a). For the inverse problem considered here, we use synthetic experimental measurements of concentrations of a pollutant collected at a position of interest over time, in order to estimate the function $C_0(t)$, which represents one of the boundary conditions, given by Eq. (3.25b).

In the solution of the inverse problem, the unknowns to be estimated are represented in Fig. 3.5 by the parameters C_{0_i}, $i = 1, 2, \ldots, N_p$, referring to a discretization of $C_0(t)$, which is equally spaced in time. Once the values of C_{0_i} are known, a linear interpolation is performed to determine $C_0^{rec}(t)$, which represents the recovered function. Therefore, a Type III Inverse Problem (IP) is considered [118] where an infinite dimensional mathematical model, is used, and a finite number of parameters is estimated. We can then define the vector of unknowns, P, as follows

$$P = \left\{C_{0_1}, C_{0_2}, C_{0_3} \ldots, C_{0_i}, \ldots, C_{0_{N_p}}\right\}^T \tag{3.36a}$$

where

$$C_{0_i} = C_0(t_i), \quad i = 1, 2, 3, \ldots, N_p \tag{3.36b}$$

$$t_i = t_0 + i\Delta t, \quad i = 1, 2, 3, \ldots, N_p \tag{3.36c}$$

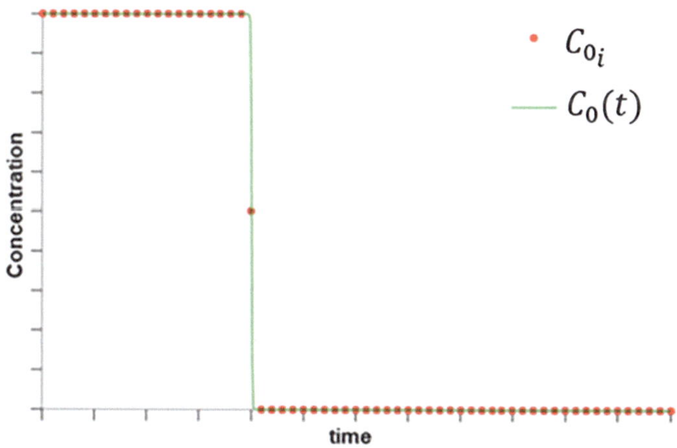

Fig. 3.5 Discretization of the function that describes the concentration at the inlet boundary of a river stretch, $C_0(t)$

$$\Delta t = \frac{t_f - t_0}{N_p} \tag{3.36d}$$

being N_p the number of unknown parameters, t_0 the initial time, $t_f = t_{Np}$, the final time, and Δt the time interval in the discretized time domain. In Eq. (3.36a) the superscript T indicates transpose.

Within the environmental context, identifying the function $C_0(t)$ means understanding how the monitored substance has been released in the hypothetical fluvial course considered here. This information is important for the management of water bodies, since it is possible to use these results in the identification of both the origin and load of possible discharges¸ according to the standards stipulated in resolutions 357/2005 and 430/2011 of the Brazilian National Environmental Council [33, 34].

3.3.1 Bayesian Inference

Methods for the solution of inverse problems based on Bayesian inference are techniques that do not usually use the inversion of matrix operators. In this way, this methodology has a certain advantage in relation to traditional inversion techniques, since these operators are subject to numerical instabilities [122].

According to Ref. [74], the objective of statistical inversion methods is to reformulate the inverse problem as a problem of statistical inference. In the Bayesian approach, attempts are made to incorporate into the model all the available information in order to reduce uncertainties about the parameters of interest. The whole theory of statistical inversion is based on the following principles: (i) all the parameters of the model are considered as random variables; (ii) randomness describes the a priori level of information; (iii) the a priori level of information is modeled with a probability distribution; (iv) the solution of the inverse problem consists on the posterior joint probability density function.

The Bayesian theorem for inverse problems comprises all the principles of statistical inversion quoted above, and is formulated as follows:

$$\pi_{pos}(\boldsymbol{P}) = \pi(\boldsymbol{P}|\boldsymbol{Y}) = \frac{\pi(\boldsymbol{Y}|\boldsymbol{P})\pi_{pr}(\boldsymbol{P})}{\pi(\boldsymbol{Y})} \tag{3.37}$$

where $\boldsymbol{P}$ is the vector of unknown parameters, $\boldsymbol{Y} = \left\{Y_1, Y_2, \ldots Y_{N_d}\right\}^T$ represents the vector of experimental measurements, $\pi_{pos}(\boldsymbol{P})$ is the distribution a posteriori, $\pi(\boldsymbol{P})$ is the likelihood, function $\pi_{pr}(\boldsymbol{P})$ is the distribution a priori and $\pi(\boldsymbol{Y})$ is the marginal probability density, which plays a role of normalization constant.

Assuming that the errors related to the experimental data are additive, and are modeled by a Gaussian distribution, with known mean and standard deviation independent of the parameters, the likelihood can be expressed as [64]:

$$\pi(\mathbf{Y}|\boldsymbol{P}) = \frac{1}{\sqrt{(2\pi)^{N_d}}}\frac{1}{\det(\boldsymbol{W})} \times \exp\left[-\frac{1}{2}(\boldsymbol{Y} - \boldsymbol{Y}_{calc}(\boldsymbol{P}))^T \boldsymbol{W}^{-1}(\boldsymbol{Y} - \boldsymbol{Y}_{calc}(\boldsymbol{P}))\right] \tag{3.38}$$

where W is the covariance matrix of the experimental data error, N_d is the number of measurements, and $\boldsymbol{Y}_{calc}$ is the vector with the values calculated by the direct problem.

In this chapter we opted for two non-informative prioris in the analysis of the inverse problem. This approach was chosen because of the lack of knowledge about how a pollutant is thrown into a river, since the only available information is the concentration measurements at a particular position downstream from the dumping site.

The priori distributions used were Total Variation (TV) and Gaussian Smoothing (GS) which, for the one-dimensional case considered in this chapter, according to Ref. [74] have their expressions given by:

$$\pi_{TV}(\boldsymbol{P}) \propto \exp[-\alpha TV(\boldsymbol{P})] \tag{3.39a}$$

where

$$TV(\boldsymbol{P}) = \sum_{i=1}^{N_p-1} \Delta t\left[\left|C_{0_{i+1}} - C_{0_i}\right|\right] \tag{3.39b}$$

and

$$\pi_{GS}(\boldsymbol{P}) \propto \exp[-\alpha GS(\boldsymbol{P})] \tag{3.40a}$$

where

$$GS(\boldsymbol{P}) = \sum_{i=1}^{N_p-1} \left(C_{0_{i+1}} - C_{0_i}\right)^2 \tag{3.40b}$$

and α is a regularization parameter.

3.3.2 Markov Chain Monte Carlo Method

The choice of a given a priori distribution is an important step in the theory of statistical inversion, since there are prioris that result in analytic a posteriori distributions, as in the case of the maximum a posteriori (MAP) function. When it is not possible to find an analytical expression for the posteriori, it is necessary to use numerical sampling methods so that an inference about the target distribution can be made from such samples.

The numerical method most used to exploit a posteriori distribution is the Monte Carlo simulation [55]. Among the sampling strategies in the literature, the Markov Chain Monte Carlo (MCMC) method is used in this chapter.

The MCMC is an iterative version of the traditional Monte Carlo methods. The idea is to generate a large sample of the posterior distribution, and then calculate sample estimates of characteristics of this distribution using iterative simulation techniques, based on Markov chains [55].

A Markov chain is a stochastic process $\{\boldsymbol{P}^0, \boldsymbol{P}^1, \ldots\}$ such that the distribution of $\boldsymbol{P}^k$, given all previous values $\boldsymbol{P}^0$, $\boldsymbol{P}^1, \ldots, \boldsymbol{P}^{k-1}$, depends only on $\boldsymbol{P}^{k-1}$ [60], that is, only the current state is necessary for estimating the future state. Mathematically, it is written

$$\pi\left(\mathbf{P}^k \in S|\mathbf{P}^0, \mathbf{P}^1, \ldots, \mathbf{P}^{k-1}\right) = \pi\left(\mathbf{P}^k \in S|\mathbf{P}^{k-1}\right) \tag{3.41}$$

for any set S.

In order to implement the MCMC method it is necessary to use a transition density, i.e. an auxiliary probability density function $q\ (\boldsymbol{P}^*|\boldsymbol{P}^k)$, which gives the possibility of generating a new solution candidate $\boldsymbol{P}^*$ from the current state of the chain, $\boldsymbol{P}^k$ [74]. The methodology chosen for the construction of the chains in this work was the Metropolis–Hastings Algorithm. The steps necessary for its implementation are described next:

1. Initialize the counter of the chain states $k = 0$, and determine an initial state for the chain $\boldsymbol{P}^0$;
2. Generate a candidate P^* from the current state of the chain $\mathbf{P}^k$ using the normal transition density $q\left(\boldsymbol{P}^*|\boldsymbol{P}^k\right) = N\left(\boldsymbol{P}^k, \sigma_q\right)$, where σ_q is the transition step;
3. Calculate the acceptance probability

$$\varepsilon = \min\left[1, \frac{\pi(\boldsymbol{Y})q\left(\boldsymbol{P}^k\right)}{\pi(\boldsymbol{Y})q(\boldsymbol{P}^*)}\right] \tag{3.42}$$

4. Generate a random number r drawn from a uniform distribution between 0 and 1 ($r \sim U(0,1)$);
5. If $r < \varepsilon$ then $\boldsymbol{P}^{k+1} = \boldsymbol{P}^*$, otherwise $\boldsymbol{P}^{k+1} = \boldsymbol{P}^k$;
6. Increase the counter $k = k + 1$ and return to step 2.

In this way, it is possible to generate a sequence that represents the posteriori distribution and the inference in this distribution is obtained by the inference in the samples $\{\boldsymbol{P}^0, \boldsymbol{P}^1, \ldots \boldsymbol{P}^{N_{MCMC}}\}$, where N_{MCMC} is the size of the string, i.e. the total number of states generated. Equation (3.42) is known as the Hastings Ratio [122], being responsible for the probability of accepting a candidate P^*.

It is important to mention that the initial states of the chain, known as heating samples or burn-in, should be discarded because the equilibrium distribution most likely may not have been reached during the first iterations [170]. In Fig. 3.6 it is observed that after the burn-in states, the chain starts to have an equilibrium behavior.

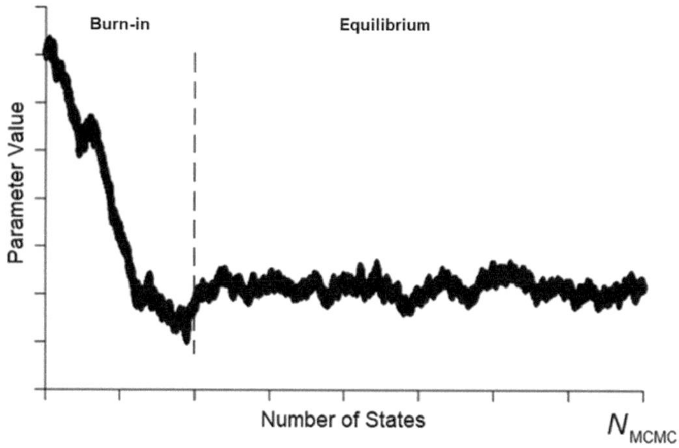

Fig. 3.6 Schematic representation of the states in the MCMC method, with the burn-in and equilibrium regions

In the present chapter, a uniform transition auxiliary probability density, q, was adopted in the Metropolis–Hastings algorithm, simplifying Eq. (3.42), since the probability of $q\,(\boldsymbol{P}^*|\boldsymbol{P}^k) = q\,(\boldsymbol{P}^k|\boldsymbol{P}^*)$. Thus, one can rewrite the Hastings Ratio as:

$$\varepsilon = \min\left[1, \frac{\pi\,(\mathbf{P}^*|\boldsymbol{Y})}{\pi\left(\mathbf{P}^k|\boldsymbol{Y}\right)}\right] \tag{3.43}$$

In order to calculate the terms in Eq. (3.43), Eq. (3.37) is used, resulting then the need to calculate only the values of the likelihood function and of the priori, for $\boldsymbol{P}^*$ and $\boldsymbol{P}^k$. Observe that the marginal probability cancels out.

3.4 Results and Discussion

In this section we present the results for the estimation of a function $C_0(t)$, which represents the boundary condition given by Eq. (3.25b), at $x = 0$, by using the inverse problem formulation described in Sect. 3.3. For this, we use synthetic experimental data calculated with the GITT and a high order of truncation, where the experimental noise is simulated through a normal distribution of zero mean and standard deviation equal to a certain percentage, for instance 2.5% or 5% of the peak concentration of the pollutant plume.

As mentioned before, besides the computation of the CI, the quality of the solution of the inverse problem, i.e. the quality of the fit between the calculated and the exact parameters will be also analyzed by means of the root mean square error. The closer to zero is the RMSE, the more accurate the adjustment is. The RMSE is given by

$$RMSE = \left[\frac{1}{N_p} \sum_{i=1}^{N_p} (C_{0_i}^{exact} - C_{0_i})^2 \right]^{\frac{1}{2}} \tag{3.44}$$

where $C_{0_i}^{exact} = C_0(t_i)$ and C_{0_i}, with $i = 1, 2, \ldots, N_p$, are the values estimated with the solution of the inverse problem.

Three different scenarios for the pollutants discharge into a river are considered to encompass situations of interest in environmental studies. The three scenarios involve time dependent pollutant discharges consisting of two non-simultaneous pulses with different intensities. In the first one both pulses are flat, while in the second both are Gaussian, and in the third a combination is considered, being one pulse Gaussian and the other one flat. Table 3.1 presents twelve cases, for the three different discharge scenarios, the two non-informative prioris (TV and GS), and two different levels of experimental noise (represented by σ_{exp} in Eq. (1.27)).

It is important to note that no analytical solutions were found in the literature for any of the launch scenarios discussed here.

The comparison of the results for the different cases studied will be analyzed by calculating the 95% confidence (credibility) interval (CI), together with the Root Mean Square Error (RMSE) and the recovered mass (if the contaminant analyzed is conservative).

The 95% confidence (credibility) interval can be calculated as follows [144]. Let $p = 0.95$ be the confidence level, N_{MCMC} the chain size and $N_{burn-in}$ the number of states

Table 3.1 Test cases considered

Cases	Scenarios	Priori[a]	σ_{exp} corresponding to (%)
1	I,$C_{0_1}(t)$	TV	5
2	I,$C_{0_1}(t)$	TV	2.5
3	I,$C_{0_1}(t)$	GS	5
4	I,$C_{0_2}(t)$	GS	2.5
5	II,$C_{0_2}(t)$	TV	5
6	II,$C_{0_2}(t)$	TV	2.5
7	II,$C_{0_2}(t)$	GS	5
8	II,$C_{0_2}(t)$	GS	2.5
9	III,$C_{0_3}(t)$	TV	5
10	III,$C_{0_3}(t)$	TV	2.5
11	III,$C_{0_3}(t)$	GS	5
12	III,$C_{0_3}(t)$	GS	2.5

I: Two flat pulses
II: Two Gaussian pulses
III: One Gaussian and one flat pulses
[a] TV: Total Variation
GS: Gaussian Smoothing

in the heating phase of the MCMC method. Once these quantities are established, the CI for each parameter can be obtained with the following steps:

1. Neglect the $N_{burn-in}$ heating samples;
2. Compute $m = \frac{(1-p)}{2}(N_{MCMC} - N_{burn-in})$, where m is the integer value of this expression;
3. Increasingly order the Markov states;
4. Neglect the first and the last m states of each chain;
5. After applying the previous steps, the lower limit of the CI will be the lowest value left in the chain and the upper limit will be the highest value left in the chain.

In order to facilitate the comparison between the recovered masses, a procedure was adopted so that they are the same in all cases. First, each of the three launch scenarios will be represented by a function, given respectively by $C_{o_1}(t)$, $C_{0_2}(t)$ and $C_{0_3}(t)$. Assuming a river with a constant volumetric flow rate Q and choosing the mass M_1 of the scenario I as base, one obtains

$$M_1 = Q \int_0^{t_f} C_{o_1}(t)dt \tag{3.45}$$

Next, the mass of another type of discharge is calculated, for example, the mass M_2 of scenario II.

$$M_2 = Q \int_0^{t_f} C_{0_2}(t)dt \tag{3.46}$$

It is now possible to find a correction factor (CF), which will be used to equalize the areas under the curves for $C_{o_1}(t)$ and $C_{o_2}(t)$ aiming at equalizing the pollutant masses in the first two scenarios,

$$CF = \frac{M_1}{M_2} \tag{3.47}$$

Finally, by calculating the resulting mass due to $C_{o_2}(t)$ using the correction factor, results

$$\begin{aligned} M_2^{CF} &= Q \int_0^{t_f} C_{o_2}(t).CFdt = Q.CF \int_0^{t_f} C_{o_2}(t)dt \\ &= \frac{M_1}{M_2} Q \int_0^{t_f} C_{o_2}(t)dt = \frac{M_1}{M_2} M_2 = M_1 \end{aligned} \tag{3.48}$$

The procedure for equalizing the mass M_3 of scenario III is analogous to that described for M_2, resulting in M_3^{CF}. For the sake of simplicity, all masses will be referenced from this point onwards as equals, thus discarding the *CF* indicator.

For the estimation of the boundary condition given by Eq. (3.25b) with the Markov Chain Monte Carlo method, we defined the chains with a total of $N_{MCMC} = 80{,}000$ states and a burn-in of $N_{burn-in} = 40{,}000$ states. We set each scenario with a total of $N_p = 61$ parameters, and all simulations were performed in a domain of $L = 400$ m, with the synthetic experimental data being taken at the position $x = 150$ m for 20 min (1200 s). The value of the mean flow velocity, the longitudinal dispersion coefficient and the river volumetric flow shall be considered constant and equal to $u = 0.59$ m s^{-1}, $D_l = 1.88\text{m}^2$ s^{-1} and $Q = 538$ L s^{-1}, respectively. These values were chosen based on the field experiment conducted by Souza (2009) on the São Pedro River, located in the neighborhood of São Pedro da Serra in the city of Nova Friburgo/RJ, Brazil [155, 164]. During the experiment, the river flow was measured using both the half section method and the chemical method [8, 14, 164].

Before starting the analysis of the test cases considered here, it is important to highlight the difficulty in estimating the last 11 parameters defined for time $t \geq 1{,}000$ s with the MCMC method. This problem was already expected since at this time the plume of contaminants could not be fully detected at the position defined as the sampling point. Thus, for the calculation of the mean square error, only the first 50 parameters were considered as a consequence of the low sensitivity in the estimation process of C_{o_i}, for $i = 51$ up to 61.

3.4.1 *Scenario I—Two Non-simultaneous Flat Pulses*

The first pollutant release scenario represents two constant discharges of finite duration with different intensities:

$$C_{0_1}(t) = \frac{C_{0_{1,1}}}{1 + \exp\exp(\gamma(t - \tau_1)) + \exp\exp(\gamma(\tau_2 - t))} + \frac{C_{0_{1,2}}}{1 + \exp(\gamma(t - \tau_3)) + \exp(\gamma(\tau_4 - t))} + C_1 \tag{3.49}$$

where τ_1 and τ_2 indicate the time interval of the first discharge with concentration $C_{0_{1,1}}$ ($\tau_1 \leq t \leq \tau_2$), as well as τ_3 and τ_4 represent the time interval of the second discharge with concentration $C_{0_{1,2}}$ ($\tau_3 \leq t \leq \tau_4$), and C_1 is the concentration of pollutant already existing in the river. The degree of smoothness of the abrupt concentration variations is controlled by the parameter γ.

In Fig. 3.7 it is shown the function to be estimated, $C_{0_1}(t)$, together with the 61 parameters, C_{0_i}, to be estimated with the solution of the inverse problem. The coefficients of Eq. (3.49) are given in Table 3.2. It is important to note that the two launches considered had a duration of 200 s each, resulting in a total pollutant mass of 16,140 g.

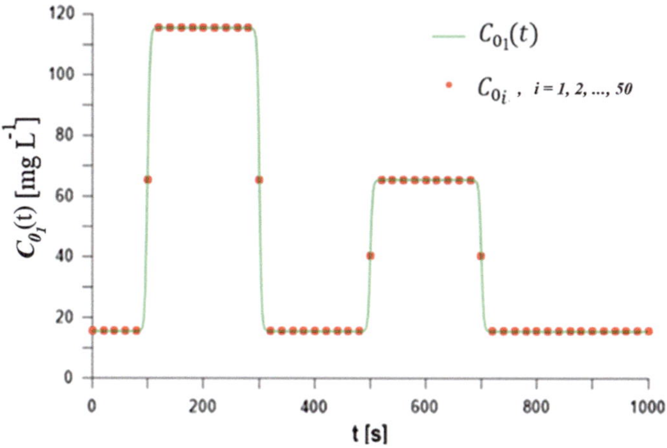

Fig. 3.7 Pollutant discharge $C_{0_1}(t)$ in Scenario I

Table 3.2 Values of the coefficients for $C_{0_1}(t)$ in Eq. (3.49)

$C_{0_{1,1}}$ (mg L^{-1})	τ_1(s)	τ_2(s)	$C_{0_{1,2}}$ (mg L^{-1})	τ_3(s)	τ_4(s)	C_1 (mg L^{-1})	γ
100	100	300	50	500	700	15.5	−0.5

In Fig. 3.8 are represented the synthetic experimental data, taken at position $x = 150$ m, using two distinct noise levels calculated from a percentage of the peak concentration of the exact data. It should be noted that, as described in Table 3.2, a concentration of $C_1 = 15.5$ mg L^{-1} of pollutant already existing in the river before the time of launching was considered.

For the first case considered, as shown in Table 3.1, the solution of the IP was obtained using a Total Variation priori with $\alpha = 0.0083$, and an experimental noise $\sigma_{\text{exp}} = 5.52$ mg L^{-1}. The recovered function, $C_{0_1}^{rec}(t)$ yields a RMSE of

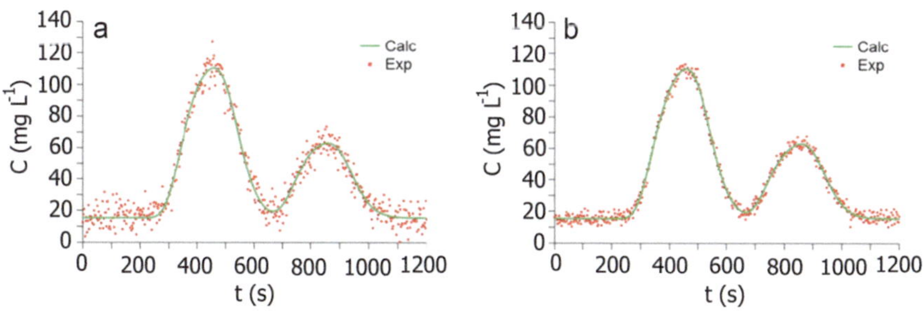

Fig. 3.8 Simulated experimental data in scenario I. Synthetic experimental data with (a) $\sigma_{\text{exp}} = 5.52$ mg L^{-1} and (b) $\sigma_{\text{exp}} = 2.76$ mg L^{-1}. See Eq. (1.27)

9.0363 mg L^{-1}, and 34 parameters (68%) were within the 95% confidence interval. The adjustment to the synthetic experimental data, given by the solution of the direct problem, substituting $C_{0_1}^{rec}(t)$ in Eq. (3.44), present a RMSE of 5.5899 mg L^{-1}, with a recovered mass of 16, 314.3 g (relative error of 1.08% with respect to the exact mass).

In the MCMC method 31,117 states were accepted, approximately 39%, with a transition density step of 0.85. Due to the number of parameters in the inverse problem, the chains and histograms were analyzed only for the parameters $C_{0_{15}}$, $C_{0_{35}}$ and $C_{0_{50}}$. For the chains relative to $C_{0_{35}}$ and $C_{0_{50}}$ one can observe a convergence with approximately 20,000 states, with the corresponding histograms close to a normal distribution. The parameter $C_{0_{15}}$ was not able to reach steady state with the number of states defined for the MCMC.

For case 2, see Table 3.1, a Total Variation priori was used in the inverse problem for the estimation of the function $C_{0_1}(t)$ with $\alpha = 0.0045$ and σ_{exp} corresponding to 2.5% of the peak concentration of the exact data. In this simulation 28 parameters are within the CI (56%), leading to a RMSE of 9.2622 mg L^{-1} for the recovered function. The adjustment between the solution of the direct problem using $C_{0_1}^{rec}(t)$ as the boundary condition of the advection–dispersion equation and the synthetic data, leads to a RMSE of 2.6536 mg L^{-1}. The mass recovered obtained was 16,182.5 g, which represents a relative error of 0.26%.

Out of the total of 80,000 samples employed in the MCMC, a total of 31,348 states were accepted when using a step of the transient density of 0.65. In this case, the amplitudes of the confidence intervals were smaller in relation to the previous case. The chains and the histograms of this case presented a behavior similar to those described in case 1.

In the third case, the estimation of $C_{0_1}(t)$ was performed considering $\alpha = 1.3$ in the solution of the inverse problem, with the Gaussian Smoothing priori and experimental data with experimental error $\sigma_{\text{exp}} = 5.52$ mg L^{-1}. The RMSE for parameter setting was 11.2514 mg L^{-1}, where 30 of them are within the CI (60%). The adjustment of the direct problem solution with respect to the experimental data with a noise corresponding to 5% of the peak concentration of the exact data showed a RMSE of 5.5539 mg L^{-1} and a recovered mass of 16,302.1 g (relative error of 1% with respect to the exact mass discharged into the river).

Out of the total of 80,000 states set in the MCMC method, 36,107 states were accepted (45.1%) with a 1.05 step in the transition density. The chains and histograms of $C_{0_{15}}$, $C_{0_{35}}$ and $C_{0_{50}}$ have reached steady state with the size of the stipulated chain, with the histograms resembling a normal distribution.

Finally, we have the results for case 4 of scenario I where the Gaussian Smoothing priori was used in the solution of the inverse problem with $\alpha = 1.35$. The adjustment of the parameters resulted in a RMSE of 11.0054 mg L^{-1}, where 27 (54%) of them had their values within the confidence interval. The solution of the direct problem using the $C_{0_1}^{rec}(t)$ function as the boundary condition of the advective-dispersion equation together with the synthetic experimental data with $\sigma_{\text{exp}} = 2.76$ mg L^{-1} presented a

Table 3.3 Summary of results for the scenario I

Case	Parameters within the CI		RMSE (mg L^{-1})		Mass (g)	
	Total number	(%)	$C_{0_1}^{rec}(t)$	Data	Recovered	Error (%)
1	34	68	9.0363	5.5899	16,314	1.08
2	28	56	9.2622	2.6536	16,183	0.26
3	30	60	11.2514	5.5539	16,302	1.00
4	27	54	11.0054	2.6597	16,181	0.25

RMSE of 2.6597 mg L^{-1} and 16,181.3 g of recovered mass, that is, a difference of 41.27 g in relation for the exact mass, this represents a relative error of 0.25%.

The MCMC method presented a total of 38,167 accepted states out of the 80,000 defined samples. The chains showed a similar convergence, where approximately in 15,000 states equilibrium has been reached.

In order to facilitate the comparison between results found for the release scenario I, the results for cases 1 to 4 will be presented in a summarized way. Analyzing the data in Table 3.3 we have that the cases where the Total Variation priori was used presented a greater number of parameters within the 95% confidence interval, in addition to lower values of RMSE for the recovered function. Another factor worth mentioning is the proximity of the recovered mass value with the exact mass for cases 2 and 4, in which experimental data with smaller errors were used.

3.4.2 *Scenario II—Two Non-simultaneous Gaussian Pulses*

The second pollutant discharge scenario, given by the function $C_{0_2}(t)$, comprises two concentration peaks given by the sum of two Gaussian distributions as follows

$$C_{0_2}(t) = \frac{C_{0_{2,1}} \times CF}{\sigma_{2,1}\sqrt{2\pi}} \exp\left(-\frac{1}{2}\frac{(t-t_1)^2}{\sigma_{2,1}^2}\right) + \frac{C_{0_{2,2}} \times CF}{\sigma_{2,2}\sqrt{2\pi}} \exp\left(-\frac{1}{2}\frac{(t-t_2)^2}{\sigma_{2,2}^2}\right) + C_1 \tag{3.50}$$

where $C_{0_{2,1}}$ and $C_{0_{2,2}}$ are the peak concentrations, t_1 and t_2 the times where the peaks of $C_{0_{2,1}}$ and $C_{0_{2,2}}$ occur respectively, $\sigma_{2,1}$ and $\sigma_{2,2}$ represent the spreading relative to the concentration (standard deviation) peaks and C_1 corresponds to the concentration of pollutant already existing in the river.

Figure 3.9 shows the function $C_{0_2}(t)$, being also represented the 50 parameters C_{0_i} used in the inverse problem. The coefficients used in Eq. (3.50), as well as the correction factor CF, see Eq. (3.47), are given in Table 3.4.

In Fig. 3.10 are represented the synthetic experimental data, taken in the same position as in scenario I, using two different noise levels calculated from a percentage of the peak concentration of the exact data. In Fig. 3.10a, it is possible to observe

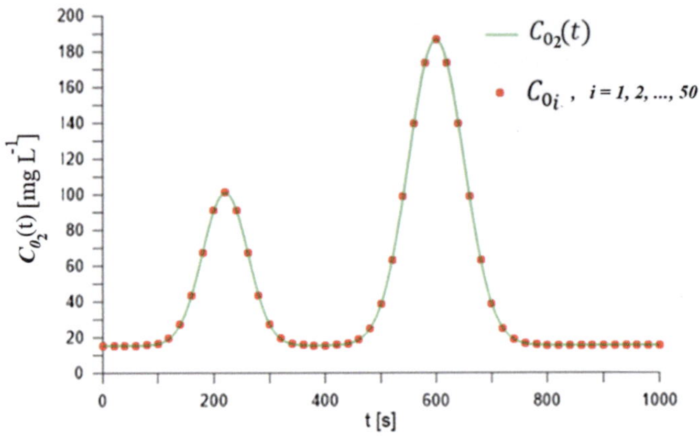

Fig. 3.9 Pollutant discharge $C_{0_2}(t)$ in Scenario II

Table 3.4 Values of the coefficients for $C_{0_2}(t)$ in Eq. (3.50)

$C_{0_{2,1}}$ (mg L^{-1})	t_1(s)	$\sigma_{2,1}$ (mg L^{-1})	$C_{0_{2,2}}$ (mg L^{-1})	t_2(s)	$\sigma_{2,2}$ (mg L^{-1})	C_1 (mg L^{-1})	$CF = \frac{M_1}{M_2}$
2000	220	40	5000	600	50	15.5	4.286

a higher experimental noise, corresponding to 5% of the maximum concentration, while in Fig. 3.10b the experimental noise is up to 2.5% of the same concentration.

As shown in Table 3.1, the simulation performed for case 5 considered a Total Variation priori using $\alpha = 0.0037$. For the recovered function a RMSE of 5.8282 mg L^{-1} was obtained, where 47 parameters (94%) were within the confidence interval. The comparison of the solution of the direct problem using the function $C_{0_2}^{rec}(t)$ with the synthetic data considering an experimental error of $\sigma_{\text{exp}} = 6.77$ mg L^{-1} resulted in a RMSE of 6.35164 mg L^{-1}, while the mass recovered was 15,978.1 g (relative error of 1% in comparison to the exact mass launched in the river).

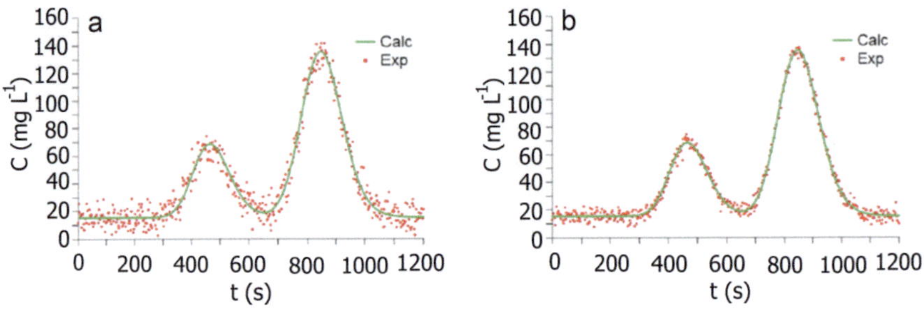

Fig. 3.10 Simulated experimental data in scenario II. Synthetic experimental data with (**a**) $\sigma_{\text{exp}} = 6.77$ mg L^{-1} and (**b**) $\sigma_{\text{exp}} = 3.37$ mg L^{-1}. See Eq. (1.27)

In the MCMC method, a total of 41,019 states were accepted using 1.05 as the value of the transition density step. The chains and the histograms of the $C_{0_{35}}$ and $C_{0_{50}}$ parameters presented good convergence with the 80,000 states defined for the MCMC, but parameter $C_{0_{15}}$ didn't.

For case 6, a Total Variation priori was used in the inverse problem for the estimation of the function $C_{0_2}(t)$ with $\alpha = 0.01$. The estimated curve in the inverse problem resulted in a RMSE of 6.1439 mg L^{-1}, where 36 parameters (72%) were within the limits of the 95% CI. The solution of the direct problem using $C_{0_2}^{rec}(t)$ as the boundary condition of the advection–dispersion equation together with the synthetic data with $\sigma_{\exp} = 3.37$ mg L^{-1} led to a RMSE of 3.1229 mg L^{-1} and the recovered mass has a value of 16,060.9 g (relative error of 0.49%).

The total number of states accepted in the MCMC method was 32,004, i.e. approximately 40%, with a pitch of 0.62 in the density of transition. The chains and the histograms for this case has shown a similar behavior to the previous case.

For the case 7 we used a Gaussian Smoothing priori in the inverse problem in order to estimate the function $C_{0_2}(t)$. The results obtained for this simulation show that for the adjustment in relation to the parameters a RMSE of 6.97958 mg L^{-1} was obtained with 45 of them, or 90%, inside the CI. The adjustment with respect to the experimental data presented a RMSE of 6.31703 mg L^{-1} with 15,993.1 g of recovered mass, corresponding to a relative error of 0.91% with respect to the mass of pollutant launched in the river.

The number of states accepted in the MCMC method application was 27,972, i.e. approximately 34.9%, with a step of 1.65 in the transition density. For the Gaussian Smoothing priori a value of α equal to 0.95 was adopted. The chain for $C_{0_{15}}$ and $C_{0_{35}}$ reached the equilibrium with less than 5,000 states, while the chain for $C_{0_{50}}$ reached equilibrium approximately with 50,000 states.

In case 8, it is considered the simulation performed with a Gaussian Smoothing priori using $\alpha = 1.2$ in the resolution of the inverse problem. For the function recovered, $C_{0_2}^{rec}(t)$ a value of 6.6001 mg L^{-1} was obtained for the RSME, where 41 parameters (82%) were within the 95% confidence interval. The adjustment with respect to the experimental data using $C_{0_2}^{rec}(t)$ in the solution of the direct problem resulted in a RMSE of 3.0361 mg L^{-1} with 16,066.7 g of recovered mass, which corresponds to a difference of 73.3 g in relation to the exact mass of the pollutant, i.e. 0.45%.

In the MCMC method application, a total of 36,444 states were accepted using 0.71 as the value of the transition density step. Good convergence was obtained for $C_{0_{35}}$, but for $C_{0_{15}}$ and $C_{0_{50}}$ the total of 80,000 states defined for the MCMC was still not sufficient for this parameter to reach the equilibrium state.

The comparison between the results of the cases studied for this pollutant launch scenario is shown in Table 3.5. It is possible to observe that cases 6 and 8, where synthetic data had a lower noise level, the recovered masses had a lower relative error percentage than cases 5 and 7. It is also important to point out that although the recovered mass is more accurate for the cases where smaller experimental errors were considered, the RMSE for the recovered function presented lower values for the cases with the higher experimental error.

Table 3.5 Summary of results for the scenario II

Case	Parameters within the CI		RMSE (mg L^{-1})		Mass (g)	
	Total number	(%)	$C_{0_2}^{rec}(t)$	Data	Recovered	Error (%)
5	47	94	5.8282	6.3516	15,978	1.00
6	36	72	6.1439	3.1229	16,061	0.49
7	45	90	6.9796	6.3170	15,993	0.91
8	41	82	6.6001	3.0361	16,067	0.45

3.4.3 Scenario III—Non-simultaneous Gaussian and Flat Pulses

Finally, there is the last analyzed pollutant release scenario which comprises a mixture of the two previous ones, that is, a function is estimated, $C_{0_3}(t)$, which is composed by a curve with the form of a normal distribution, followed by a finite constant launch pulse during a given time interval,

$$C_{0_3}(t) = \frac{C_{0_{3,1}} \times CF}{\sigma_{3,1}\sqrt{2\pi}} \exp\left(-\frac{1}{2}\frac{(t-t_1)^2}{\sigma_{3,2}^2}\right) + \frac{C_{0_{3,2}} \times CF}{1 + \exp\exp(\gamma(t-\tau_1)) + \exp\exp(\gamma(\tau_2 - t))} + C_1 \tag{3.51}$$

where $C_{0_{3,1}}$ is the peak concentration for the Gaussian shaped launch centered at t_1, and with a standard deviation $\sigma_{3,1}$, $C_{0_{3,2}}$ is the constant concentration for the flat pulse launch in the time interval from τ_1 to τ_2, i.e. $\tau_1 < t < \tau_2$, γ is the parameter representing the degree of smoothness of the concentration variation in the lower and upper times of the flat pulse, and C_1 is the concentration of pollutant already existing in the river.

In Fig. 3.11 it is shown the function $C_{0_3}(t)$, along with the 50 parameters, C_{0_i}, used in the inverse problem. The coefficients of Eq. (3.51) are given in Table 3.6. The two pollutant launches, Gaussian and flat pulses, have different maximum values, with the peak, $C_{0_{3,1}}$ being approximately half of $C_{0_{3,2}}$.

The experimental synthetic data, taken at $x = 150$ m, using different noise levels calculated as a percentage (5% or 2.5%) of the maximum concentration values of the exact data are shown in Fig. 3.12. The red dots represent the experimental measurements while the green line corresponds to the exact data calculated with the GITT with a high order of truncation $N_{tr} = 500$ terms.

The launch for the scenario III analysis comprises cases 9 to 12, where the ability of the proposed methodology to recover the function $C_{0_3}(t)$ is investigated using two non-informative prioris and synthetic experimental data sets with different experimental noise levels.

The results obtained for case 9 consider a Total Variation priori with $\alpha = 0.0065$, and $\sigma_{\exp}$ corresponding to an error of 5% of the peak concentration. The parameter

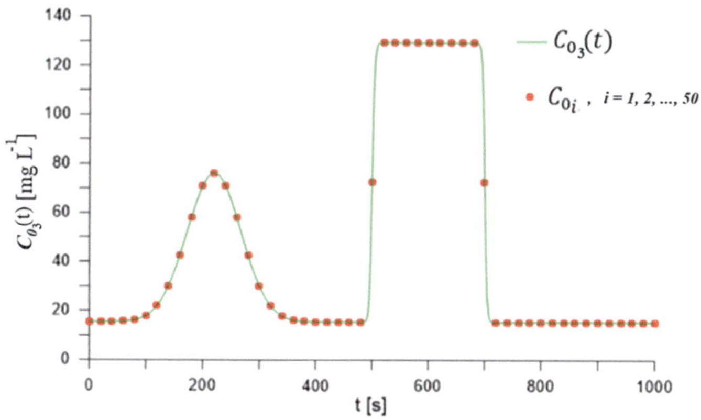

Fig. 3.11 Pollutant discharge $C_{0_3}(t)$ in Scenario III

Table 3.6 Values of the coefficients for $C_{0_3}(t)$ in Eq. (3.51)

$C_{0_{3,1}}$ (mg L^{-1})	t_1(s)	$\sigma_{3,1}$ (mg L^{-1})	$\sigma_{3,2}$ (mg L^{-1})	$C_{0_{3,2}}$ (mgL^{-1})	τ_1(s)	τ_2(s)	C_1 (mg L^{-1})	γ	$CF = \frac{M_1}{M_3}$
2000	220	30	50	50	500	700	15.5	−0.5	2.279

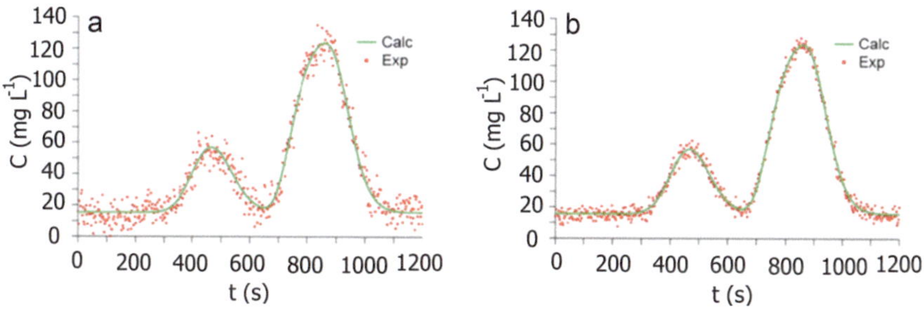

Fig. 3.12 Simulated experimental data in scenario III. Synthetic experimental data with (**a**) $\sigma_{\text{exp}} = 6.18$ mg L^{-1} and (**b**) $\sigma_{\text{exp}} = 3.09$ mg L^{-1}. See Eq. (1.27)

adjustment showed a RMSE of 7.0341 mg L^{-1}, with 42 (84%) of the parameters within the CI. The adjustment of the solution of the direct problem presented a value of 6.0759 mg L^{-1} for the RMSE, while the mass recovered was 16,035.5 g, that is, a difference of 104.5 g (0.6%) in relation to the exact mass thrown into the river. The MCMC method had a total of 37,974 accepted states, i.e. approximately 47% of the samples. The transition density step chosen was 0.85. It was observed that $C_{0_{15}}$ and $C_{0_{50}}$ were able to reach steady state before the chain for $C_{0_{35}}$, and these parameters showed unimodal histograms approaching a normal distribution.

For case 10 a Total Variation priori was also used, but now the analysis was performed considering an experimental error of 2.5% of the value of the maximum calculated concentration with $\alpha = 0.006$. A RMSE of 8.4509 mg L^{-1} was obtained in the parameter adjustment, where 38 (76%) of them were within the CI. The RMSE found was 3.0305 mg L^{-1} with 15,937.1 g of recovered mass, a difference of 202.9 g with respect to the exact mass, corresponding to a 1.26% relative error. In the MCMC method application, out of the 80,000 samples used, 34,652 states were accepted using 0.62 as the size of the transition density step. With regard to the $C_{0_{15}}$, $C_{0_{35}}$ and $C_{0_{50}}$ chains, similar behaviors were obtained as in the previous case, although the amplitudes of the confidence intervals found were narrower.

For case 11, a Gaussian Smoothing priori was used in the estimation of the function $C_{0_3}(t)$ with $\sigma_{\text{exp}} = 6.18$ mg L^{-1}. For the adjustment in relation to the parameters a RMSE of 11.0239 mg L^{-1} was obtained with 34 (68%) of them within the CI. A RMSE of 6.0455 mg L^{-1} was obtained with 16,012.1 g of recovered mass corresponding to a deviation of 127.9 g with respect to the exact mass launched into the river, which corresponds to a 0.79% error. In the MCMC method, 37,503 states were accepted, the step in the transition density being 1.15. Good convergence was obtained for the parameters $C_{0_{15}}$, $C_{0_{35}}$ and $C_{0_{50}}$ chains, as well as a behavior approaching a normal distribution for the histograms.

Finally, we have the results for case 12 where a Gaussian Smoothing priori was used in the solution of the inverse problem to determine the function $C_{0_3}(t)$, setting $\alpha = 1.4$. In the adjustment of the parameters a RMSE of 11.2634 mg L^{-1} was obtained, where 29 (58%) of them had their values within the confidence interval. The adjustment in relation to the experimental data with the use of the recovered function $C_{0_3}^{rec}(t)$ in the solution of the direct problem presented a RMSE of 3.01572 mg L^{-1}, with 15,944.1 g of mass recovered, corresponding to a relative error of 1.21% compared to the exact mass of the pollutant launched in the river. The number of states accepted in the MCMC application was 21,552, i.e. approximately 27%, with a step of 0.97 at the transition density. The chains for the $C_{0_{35}}$ and $C_{0_{50}}$ parameters were able to reach steady state, and their histograms show a behavior similar to that of a normal one, but for the $C_{0_{15}}$ didn't.

In order to synthesize the results obtained for the launch scenario III, Table 3.7 summarizes the RMSE values, the number of parameters within the 95% confidence interval, in addition to the amount of mass recovered, and its relative error for cases 9 to 12. It is important to note that in cases 9 and 10, where the Total Variation priori was used, were obtained lower values of the RMSE for the recovered function, $C_{0_3}^{rec}(t)$, in relation to cases 11 and 12, in which was used a Gaussian Smoothing priori. In addition, still analyzing the priori, Total Variation was also able to maintain a greater number of parameters within the confidence interval. Another detail that deserves attention is the fact that the recovered masses for cases 9 and 11, which used the synthetic data with the largest experimental errors, presented a lower relative error than cases 10 and 12, in which a smaller experimental error was considered.

Table 3.7 Summary of results for the scenario III

Case	Parameters within the CI		RMSE (mgL^{-1})		Mass (g)	
	Total number	(%)	$C_{0_3}^{rec}(t)$	Data	Recovered	Error (%)
9	42	84	7.0341	6.0759	16,036	0.60
10	38	76	8.4509	3.0305	15,937	1.26
11	34	68	11.0239	6.0455	16,012	0.79
12	29	58	11.2634	3.0157	15,944	1.21

3.5 Remarks

This chapter has as its main objective the estimation of a function $C_0(t)$, which represents one of the boundary conditions of the one-dimensional advection–dispersion equation in a transient regime. For this, the Generalized Integral Transform Technique was used to solve the direct problem and the Markov Chain Monte Carlo (MCMC) method in the inverse problem formulation and solution using two non-informative prioris. The proposed methodology was able to satisfactorily recover the way a pollutant is launched in the modelled water body, even with the low levels of information employed in the priors.

The use of fifty terms, i.e. $N_{tr} = 50$, in the GITT was chosen due to the high computational cost required for the solution of the inverse problem, i.e. approximately 62 h for each case using a computer with Intel Core i5 processor, 2.67 GHz and 4 GB of RAM. In addition, this order of truncation was the one that presented the minimum convergence able to reach good results in the estimation of $C_0^{rec}(t)$ for the cases in which lower experimental errors were considered.

Regarding the methodologies chosen for the non-informative a priori information, the RMSE of the recovered function presented lower values using Total Variation. Only in case 8 the Gaussian Smoothing priori obtained a better RMSE value compared to case 6 in which the same launch scenario was analyzed using the TV. Due to the stochastic characteristics of the MCMC, a more complete analysis of which of the two prioris, TV or GS, is best in solving the inverse problem is only possible through the use of non-parametric methods, which is beyond the scope of this book. Considering only a simulation for each case, the Total Variation priori was better in the estimation of $C_0(t)$, i.e. $C_0^{rec}(t)$.

Another interesting result is the relative error of the recovered mass. These values were very close to the exact quantities, calculated with the flow rate of 538 L s^{-1}, for all analyzed cases. This reflects the ability of the solution strategy, with GITT for the direct problem and MCMC for the inverse problem, to satisfactorily estimate the pollutant mass, even with the difficulty of convergence existing for $C_{0_{51}}$ up to $C_{0_{61}}$ in scenario I.

The number of parameters within the confidence interval in cases 1 to 4 were those that presented the worst results. This behavior is due to the fact that scenario I consists of two flat pulse launches with finite duration that comprise abrupt variations in the concentrations in four different time instants. In the cases of scenario 3, which contains instead of two, one flat pulse launch, a larger number of parameters within the confidence interval is observed. Scenario 2, which is formed by two Gaussian shaped launches, did not present such abrupt variations, reaching 47 out of the 50 parameters within the CI limits. Moreover, the experimental error also has a significant influence on this number, since more intense deviations lead to broader confidence intervals and, consequently, greater uncertainties.

The difficulty of convergence found for $C_{0_{51}}$ up to $C_{0_{61}}$ in the inverse problem is due to the fact that the plume of contaminants in the three launch scenarios has not been able to reach the times where these parameters were defined, thus making the estimation process harder. The total number of states $N_{MCMC} = 80{,}000$ samples defined in the MCMC, for most cases, was not sufficient to achieve the chains equilibrium. It should be emphasized that simulations with a larger number of states were not performed due to the extremely high computational cost.

Chapter 4
Estimation of Roughness Mean Height and Turbulent Diffusion Coefficient in Estuaries

This chapter presents estimates for the roughness mean height and turbulent diffusion coefficient in a 2-D Horizontal transport model, which was applied in the estuarine region of the Macaé River, which is located at the east coast of Brazil. The estimates are obtained with computational intelligence strategies, applying stochastic optimization methods, more specifically Genetic Algorithms (GA), Particle Collision Algorithm (PCA), Luus-Jaakola (LJ), and Particle Swarm Optimization (PSO), as well as Artificial Neural Networks (ANN).

4.1 Introduction

As previously discussed in Chap. 1, historically, the evaluation of the behavior of dissolved and suspended constituents in rivers, channels and estuaries began with the development of "black box" models, in which the removal or production of the constituent was evaluated by the mass balance between its entrance and its exit in a given stretch of river or in an estuary. Such approach is based on intensive monitoring, with obvious logistical and economic constraints.

As already shown in the previous chapters, the behavior of substances in bodies of water can be estimated through mathematical models [1, 14]. In the equations that model the transport of substances in a water medium there is a number of parameters that are not always calculated directly. Therefore, their estimation is often of crucial importance to the accuracy of the simulation. The definition of the best values for these parameters can be carried out through the search, by trial and error, which can consume considerable effort and, worse, generate values that, although allowing the

Edgar Barbosa Lima, from IFRJ, Rio de Janeiro Federal Institute, co-authored this chapter.

J. Lugon Junior et al., *Inverse Problems in the Environmental Modelling of Rivers and Estuaries*, Studies in Systems, Decision and Control 246,
https://doi.org/10.1007/978-3-032-09937-2_4

simulation to a specific situation, do not extend it to situations when the model can be calibrated but not validated [79, 119, 146, 147].

The design of models to solve increasingly complex problems is a motivation in algorithmic development. Great successes have been achieved through the modeling based on biological and intelligence observation in nature. These intelligent algorithms include Artificial Neural Networks (ANN), and Genetic Algorithms (GA) among others [48, 150].

In this chapter, a strategy of automatic hydrodynamic parametrization is proposed, through the solution of inverse problems, for a numerical model used in the simulation of the transport of conservative substances in estuaries. As part of this investigation, an application is made to the estuary of the Macaé river, located at the north coast of the state of Rio de Janeiro, in Brazil. The model presented here was developed with the MOHID platform (MOdelagem HIDrodinâmica—Water Modeling System) developed by researchers of the University of Lisbon Superior Technical Institute. The calibration of the applied model was carried out using several computational intelligence strategies, such as stochastic optimization methods and Artificial Neural Networks (ANN).

4.2 Formulation and Solution of the Direct Problem

4.2.1 Formulation of the Direct Problem

The mathematical formulation used in the simulation of substance transport in estuaries usually contemplates spatial and temporal variations. In terms of space, such formulation can be three, two or one-dimensional, depending on the physical and morphological characteristics of the system. The estuarine flow is essentially three-dimensional, depending on the stratification that eventually forms in the water column, with fresh (less dense) water flowing over (denser) saline water.

To model an estuary mathematically, one must start with the Navier–Stokes equations, generalized to a geophysical fluid [112–114].[1]

$$\begin{aligned}\frac{\partial u}{\partial t}+\frac{\partial (uu)}{\partial x}+\frac{\partial (uv)}{\partial y}+\frac{\partial (uw)}{\partial z}-fv\\=-\frac{1}{\rho}\left(\frac{\partial P}{\partial x}+\frac{\partial \tau_{xx}}{\partial x}+\frac{\partial \tau_{xy}}{\partial y}+\frac{\partial \tau_{xz}}{\partial z}\right)\end{aligned} \tag{4.1}$$

[1] The derivation of the equations presented in this chapter for the direct problem formulation follows closely the one performed in Refs. [112, 113].

$$\frac{\partial v}{\partial t}+\frac{\partial (vu)}{\partial x}+\frac{\partial (vv)}{\partial y}+\frac{\partial (vw)}{\partial z}+fu$$
$$=-\frac{1}{\rho}\left(\frac{\partial P}{\partial y}+\frac{\partial \tau_{yx}}{\partial x}+\frac{\partial \tau_{yy}}{\partial y}+\frac{\partial \tau_{yz}}{\partial z}\right) \tag{4.2}$$

$$\frac{\partial w}{\partial t}+\frac{\partial (wu)}{\partial x}+\frac{\partial (wv)}{\partial y}+\frac{\partial (ww)}{\partial z}+g$$
$$=-\frac{1}{\rho}\left(\frac{\partial P}{\partial z}+\frac{\partial \tau_{zx}}{\partial x}+\frac{\partial \tau_{zy}}{\partial y}+\frac{\partial \tau_{zz}}{\partial z}\right) \tag{4.3}$$

where u, v and w are the velocity components in the x, y and z axes respectively, f is the Coriolis parameter, P is the pressure, ρ is the density of the fluid, g is the gravitational acceleration and τ_{xx}, τ_{xy}, τ_{xz}, τ_{yx}, τ_{yy}, τ_{yz}, τ_{zx}, τ_{zy} and τ_{zz} are the Reynolds stress tensor components. In addition, the effects of turbulent velocity fluctuations were parametrized semi-empirically by [114]:

$$\tau_{xx}=-\rho\left(\overline{u'u'}\right)=-\mu_x\frac{\partial u}{\partial x};\ \tau_{yx}=-\rho\left(\overline{v'u'}\right)=-\mu_x\frac{\partial v}{\partial x} \tag{4.4a, b}$$

$$\tau_{xy}=-\rho\left(\overline{u'v'}\right)=-\mu_y\frac{\partial u}{\partial y};\ \tau_{yy}=-\rho\left(\overline{v'v'}\right)=-\mu_y\frac{\partial v}{\partial y} \tag{4.5a, b}$$

$$\tau_{xz}=-\rho\left(\overline{u'w'}\right)=-\mu_z\frac{\partial u}{\partial z};\ \tau_{yz}=-\rho\left(\overline{v'w'}\right)=-\mu_z\frac{\partial v}{\partial z} \tag{4.6a, b}$$

where μ_x, μ_y and μ_z represent the turbulent viscosity coefficients, u', v' and w' are the velocity components fluctuation with time, and the upper bar indicates the time average.

Assuming that μ_x, μ_y and μ_z do not vary with x, y and z, respectively, and considering the coefficients of turbulent diffusion in each direction,

$$K_x=\frac{\mu_x}{\rho},\ K_y=\frac{\mu_y}{\rho}\text{ and }K_z=\frac{\mu_z}{\rho} \tag{4.7a, b, c}$$

Eqs. (4.1) and (4.2) can be rewritten as

$$\frac{\partial u}{\partial t}+\frac{\partial (uu)}{\partial x}+\frac{\partial (uv)}{\partial y}+\frac{\partial (uw)}{\partial z}-fv=$$
$$-\frac{1}{\rho}\frac{\partial P}{\partial x}+K_x\frac{\partial^2 u}{\partial x^2}+K_y\frac{\partial^2 u}{\partial y^2}+K_z\frac{\partial^2 u}{\partial z^2} \tag{4.8}$$

$$\frac{\partial v}{\partial t}+\frac{\partial (vu)}{\partial x}+\frac{\partial (vv)}{\partial y}+\frac{\partial (vw)}{\partial z}+fu=$$
$$-\frac{1}{\rho}\frac{\partial P}{\partial y}+K_x\frac{\partial^2 v}{\partial x^2}+K_y\frac{\partial^2 v}{\partial y^2}+K_z\frac{\partial^2 v}{\partial z^2} \tag{4.9}$$

Estuaries that are subjected to a sufficiently large tidal amplitude (more than 4 m), and with a flow of the main water supply sufficiently small, can have their hydrodynamics conveniently simulated by the system of equations that are called "Shallow Water Equations (SWE)" [45, 114, 137], which, in synthesis, considers the estuary to be homogeneous vertically. Considering that the vertical changes in momentum are negligible, the integration of Eqs. (4.8) and (4.9) within the limits between the estuary bed, $-H_0(x, y)$, and the free surface, $\eta(x, y)$, is given by:

$$\int_{-H_0}^{\eta} \frac{\partial u}{\partial t} dz + \int_{-H_0}^{\eta} \frac{\partial (uu)}{\partial x} dz + \int_{-H_0}^{\eta} \frac{\partial (uv)}{\partial y} dz + \int_{-H_0}^{\eta} \frac{\partial (uw)}{\partial z} dz - \int_{-H_0}^{\eta} fv dz = - \int_{-H_0}^{\eta} \frac{1}{\rho} \frac{\partial P}{\partial x} dz + \int_{-H_0}^{\eta} \left(K_x \frac{\partial^2 u}{\partial x^2} + K_y \frac{\partial^2 u}{\partial y^2} \right) dz + \int_{-H_0}^{\eta} K_z \frac{\partial^2 u}{\partial z^2} dz \tag{4.10}$$

$$\int_{-H_0}^{\eta} \frac{\partial v}{\partial t} dz + \int_{-H_0}^{\eta} \frac{\partial (vu)}{\partial x} dz + \int_{-H_0}^{\eta} \frac{\partial (vv)}{\partial y} dz + \int_{-H_0}^{\eta} \frac{\partial (vw)}{\partial z} dz + \int_{-H_0}^{\eta} fu dz = - \int_{-H_0}^{\eta} \frac{1}{\rho} \frac{\partial P}{\partial y} dz + \int_{-H_0}^{\eta} \left(K_x \frac{\partial^2 v}{\partial x^2} + K_y \frac{\partial^2 v}{\partial y^2} \right) dz + \int_{-H_0}^{\eta} K_z \frac{\partial^2 v}{\partial z^2} dz \tag{4.11}$$

Considering a vertically homogeneous estuary, and assuming that the velocity components (u and v) are averages of the velocities in the water column, and $-(uw)|_{\eta} + (uw)|_{-H_0} = 0$, Eq. (4.10) can be rewritten as

$$\frac{\partial (uh)}{\partial t} + \frac{\partial (uuh)}{\partial x} + \frac{\partial (uvh)}{\partial y} - \int_{-H_0}^{\eta} fv dz = - \frac{h}{\rho} \frac{\partial P}{\partial x} + \int_{-H_0}^{\eta} \left(K_x \frac{\partial^2 u}{\partial x^2} + K_y \frac{\partial^2 u}{\partial y^2} \right) dz + \int_{-H_0}^{\eta} K_z \frac{\partial^2 u}{\partial z^2} dz \tag{4.12}$$

where $h(x, y) = \eta(x, y) + H_0(x, y)$, see Fig. 4.1.

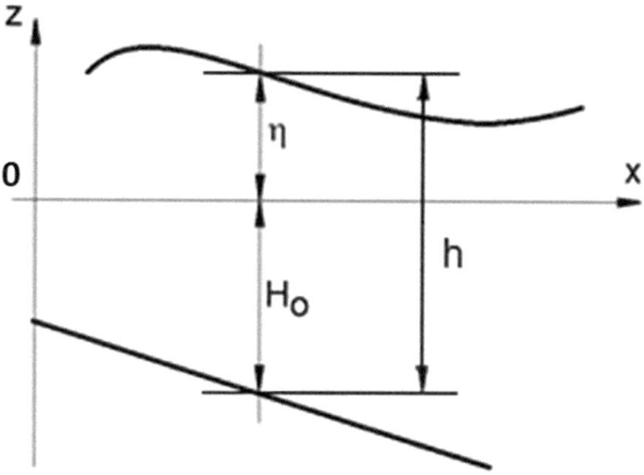

Fig. 4.1 Schematic representation of the water column

Taking into account that there is no diffusive flux of the momentum in the bed and free surface, from the second term on the right side of Eq. (3.12), results

$$\int_{-H_0}^{\eta}\left(K_x\frac{\partial^2 u}{\partial x^2}+K_y\frac{\partial^2 u}{\partial y^2}\right)dz=\frac{\partial}{\partial x}\left(K_x h\frac{\partial u}{\partial x}\right)+\frac{\partial}{\partial y}\left(K_y h\frac{\partial u}{\partial y}\right) \tag{4.13}$$

Considering that the viscous frictional integral, i.e. the last part of Eq. (4.12), can be interpreted as the difference between the components along the x-axis of the shear stresses due to wind (τ_{Wx}) and bottom friction (τ_{Bx}), that is

$$\int_{-H_0}^{\eta} K_z\frac{\partial^2 u}{\partial z^2}dz=-\frac{1}{\rho}\left(\tau_{W_x}-\tau_{B_x}\right) \tag{4.14}$$

Substituting Eqs. (4.13) and (4.14) into Eq. (4.12), we obtain:

$$\begin{aligned}\frac{\partial(uh)}{\partial t}+\frac{\partial(uuh)}{\partial x}+\frac{\partial(uvh)}{\partial y}-hfv=-\frac{h}{\rho}\frac{\partial P}{\partial x}\\+\frac{\partial}{\partial x}\left(K_x h\frac{\partial u}{\partial x}\right)+\frac{\partial}{\partial y}\left(K_y h\frac{\partial u}{\partial y}\right)-\frac{1}{\rho}\left(\tau_{W_x}-\tau_{B_x}\right)\end{aligned} \tag{4.15}$$

Using the equation of continuity

$$\frac{\partial(uh)}{\partial x}+\frac{\partial(vh)}{\partial y}+\frac{\partial\eta}{\partial t}=0 \tag{4.16}$$

in Eq. (4.15), we obtain

$$\frac{\partial u}{\partial t}+u\frac{\partial u}{\partial x}+v\frac{\partial u}{\partial y}-fv=-\frac{1}{\rho}\frac{\partial P}{\partial x}$$
$$+\frac{1}{h}\left[\frac{\partial}{\partial x}\left(K_x h\frac{\partial u}{\partial x}\right)+\frac{\partial}{\partial y}\left(K_y h\frac{\partial u}{\partial y}\right)\right]-\frac{1}{\rho h}\left(\tau_{W_x}-\tau_{B_x}\right) \tag{4.17}$$

Applying a similar derivation, but starting with Eq. (4.11), results

$$\frac{\partial v}{\partial t}+u\frac{\partial v}{\partial x}+v\frac{\partial v}{\partial y}+fu=-\frac{1}{\rho}\frac{\partial P}{\partial y}$$
$$+\frac{1}{h}\left[\frac{\partial}{\partial y}\left(K_x h\frac{\partial v}{\partial y}\right)+\frac{\partial}{\partial y}\left(K_y h\frac{\partial v}{\partial y}\right)\right]-\frac{1}{\rho h}\left(\tau_{W_y}-\tau_{B_y}\right) \tag{4.18}$$

where τ_{W_y} and τ_{B_y} are the components along the y axis of the shear stresses due to wind and bottom friction, respectively.

Considering that the vertical velocity is zero, we can simplify Eq. (4.3) by

$$\frac{\partial P}{\partial z}+\rho g=0 \tag{4.19}$$

Considering that the atmospheric pressure (P_{atm}) is subtracted from P, and the density ρ is divided by a constant density reference ρ_0, and a deviation ρ' from the constant density reference is considered (Bousinesq decomposition), after integration from the free surface to the depth z, where the pressure is calculated, we arrive at [62, 111]

$$P(z)=P_{atm}+g\rho_0(h-z)+g\int_{\eta}^{z}\rho' dz \tag{4.20}$$

The terms in Eq. (4.20) correspond to pressure at any depth, atmospheric pressure at sea surface, sea level effect and anomalous pressure integrated between that level and the surface. By using this expression and the hydrostatic approximation, the horizontal pressure gradient in the x_i direction can be divided into three contributions

$$\frac{\partial P(z)}{\partial x_i}=\frac{\partial P_{atm}}{\partial x_i}-g\rho_0\frac{\partial \eta}{\partial x_i}-g\int_{\eta}^{z}\eta\frac{\partial \rho'}{\partial x_i}dz \tag{4.21}$$

where x_i represents the spatial coordinates x and y, for $i=1$ and 2, respectively.

The total pressure gradient is the sum of the effects due to atmospheric pressure gradients, sea surface elevation (barotropic pressure gradient) and density distribution (baroclinic pressure gradient). This decomposition of pressure gradient is used in Eqs. (4.17) and (4.18).

The transport of solute and constituents in suspension can be described by the 2D-horizontal version of the advection–diffusion equation, expressed by [114]

$$\frac{\partial C}{\partial t} = u\frac{\partial C}{\partial x} + v\frac{\partial C}{\partial y} + \frac{1}{h}\left[\frac{\partial}{\partial x}\left(D_x^* h\frac{\partial C}{\partial x}\right) + \frac{\partial}{\partial y}\left(D_y^* h\frac{\partial C}{\partial y}\right)\right] + S_C - R_C \quad (4.22)$$

where C is the concentration of solute or constituents in suspension, D_x^* and D_y^* are the substance turbulent diffusion coefficients in the x and y directions, respectively, and S_C and R_C assembles all internal reactions and external inputs that can change the concentration of the constituent that is simulated. Several numerical solutions have been proposed to solve this set of four equations, (4.17), (4.18), (4.21) and (4.22), and four unknowns (h, u, v and C).

Observe that K and D^* have the following correlation [194]

$$D^* = \frac{K}{Sc_t} \quad (4.23)$$

where Sc_t is the turbulent Schmidt number for mass transport, and usually has the value very close to 1 (one). Thus, from this point on, in this chapter, these two coefficients, K and D^*, will simply be represented by D^*.

4.2.2 Numerical Solution

The direct problem presented in the previous subsection is here numerically solved using the Water module of the MOHID platform. This platform is programmed in ANSI FORTRAN 95, and the philosophy used in its construction allows the use of the model in one, two and three-dimensional approaches [115]. It is composed by more than 40 modules, and each one of them is responsible for managing certain specific type of information and/or physical model. Another important feature of MOHID is the ability to run sub-models, that allows the user to study smaller restricted areas of the larger studied region, using the boundary conditions provided by the upper level "parent" model.

The MOHID platform uses the Finite Volume Method [31, 112, 113] in the discretization of the equations. In this method, the discrete form of the equations is applied macroscopically in a control volume, and the procedure for solving the equations is independent of the geometry of each cell. This method is briefly described and used in Chap. 2 of this book.

Because volumes may vary in the course of calculations, the geometry is updated with each time step in which the physical variables are computed. In addition, the spatial coordinates are equally independent, and any geometry can be chosen for each of the dimensions. Cartesian and curvilinear coordinates can be used in the horizontal plane, while a generic vertical coordinate with different sub-domains can be used. This vertical coordinate allows the minimization of errors that are present in many classic vertical coordinates (Cartesian, Sigma, Isopycnal) as shown in Ref. [113].

Only one degree of freedom is allowed, and the grid is orthogonal Cartesian in the horizontal plane, being shifted [10], so that the horizontal velocities are solved in the center of the west faces (component u of the velocity) and south (component v of the velocity), while free surface elevation, turbulence parameters, and substance (tracer or pollutant) concentration are resolved at the center of the volume. It is also shifted in the vertical direction with the component w of the velocity, i.e. substance concentration and turbulence parameters are positioned at the bottom and top faces, while the horizontal components of velocity and elevation of the free surface are positioned at the center of the volume.

Time discretization is implemented through the semi-implicit algorithm ADI (Alternate Direction Implicit), introduced by Peaceman and Rachford in 1955 [54]. This algorithm alternately computes a horizontal velocity component in the implicit form, while computing the other component in the explicit form. This allows preserving the stability advantages, guaranteed by the implicit methods, without the disadvantage of high computational costs and associated errors.

The MOHID platform has been applied in several coastal and estuarine regions, showing itself capable of simulating complex features present in flows observed in these regions. The model has been applied both as a research tool and in consulting projects. Along the Portuguese coast, for example, different environments have been studied with MOHID, including the main estuaries (Minho, Lima, Douro, Mondego, Tejo, Sado, Mira, Arade and Guadiana) and coastal lagoons (Ria de Aveiro and Ria Formosa), [73, 114]. The model was also implemented in most Galician lagoons as Ría de Vigo [116, 117, 173, 174], as well as other lagoons [128]. Also in Brazil, a number of studies have been conducted [58, 95, 98, 101, 123, 177–179, 183].

4.2.3 Results of the Direct Problem Modeling

Hydrodynamic and transport models, built in the MOHID platform, were applied in the estuary of the Macaé River, in the coast of the State of Rio de Janeiro, Brazil. See Fig. 4.2. See also Figs. 2.1 and 2.2.

The modeled domain extends for 20 km along the main course of the river close to the mouth of the estuary, on the coast. A spatial discretization of 20 m, based on quadratic cells was adopted. The bathymetric data from the coast region were taken from the nautical chart 1507, published by the Brazilian Navy in 1974 [22], while the bathymetry of the upper region was obtained from Ref. [7]. Figure 4.3 shows the discretized domain and the bathymetry used.

Two boundary conditions were prescribed for the hydrodynamic model: (i) at the river bed, a river volumetric flow rate of 7.8 m^3s^{-1}, typical of the dry season to the Macaé river near the estuarine region [9]; and (ii) at the maritime border an astronomical tide was simulated, with 17 components [23].

The value adopted for the salinity in the fluvial region was 0.037 psu (practical salinity unit), and in the marine region 36 psu. As the initial condition, a salinity of 20 psu was defined for the entire domain, with a configuration of velocities initially null, in the hydrodynamic model. The results obtained in this model correspond to the stabilization of the saline distribution, whose results are presented in Refs. [96, 135].

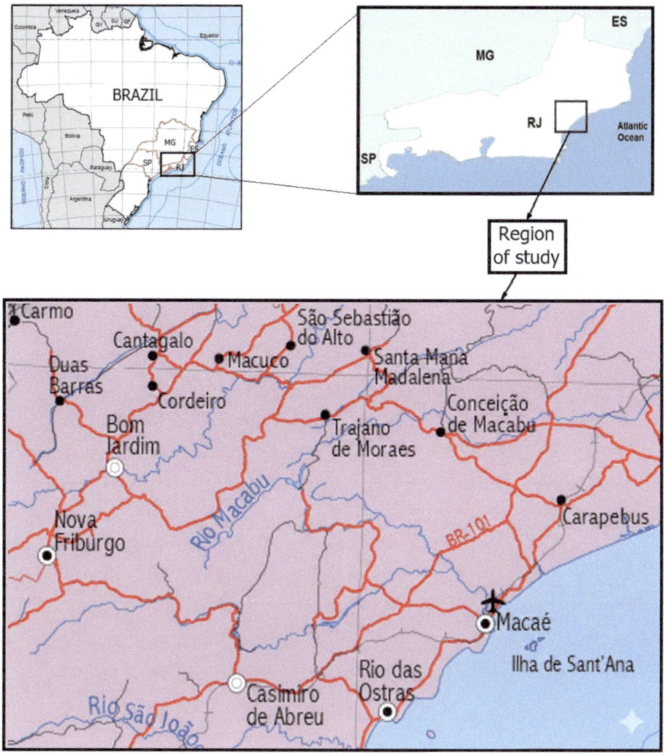

Fig. 4.2 Location of the estuary and details of the simulated region in this study. *Source* Ministry of Transport "http://www.transportes.gov.br/"

The solution of Eqs. (4.17), (4.18) and (4.21), which compose the hydrodynamic model, describes the time variation of the velocity field (u and v) and of the free surface level (η). The knowledge of the distribution of these variables is important in many respects, such as erosion and deposition zones, as well as potential flooding areas. Furthermore, the advective transport of any substance is governed by the velocity field and, finally, the volume of the estuary itself can be dynamically calculated, with obvious implications for the dilution of substances reaching the body of water. The hydrodynamic model is generally calibrated and validated from data on the free surface.

In order to evaluate the synoptic behavior of the hydrodynamic variables, two virtual monitoring stations were positioned in the estuary, as shown in Fig. 4.4a. The influence of the tidal variation at these two points close to the estuary were calculated, therefore, at 0.5 km and at 9.51 km from the mouth of the river. In

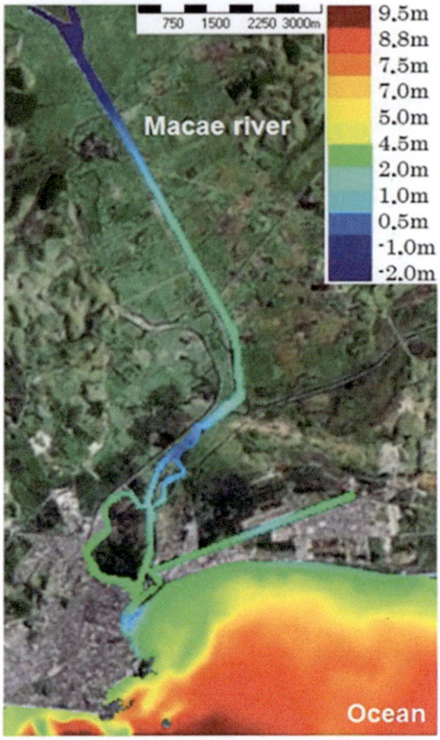

Fig. 4.3 Simulated domain bathymetry

these points, the synoptic variation of the free surface level, and the longitudinal and transversal components of the velocity, were calculated with MOHID platform, and the results are shown in Fig. 4.4b.

4.2.4 Sensitivity Analysis and Design of Experiment

A better simulation is obtained by adjusting parameters in the equations that model the studied phenomena, which corresponds to the important phase of model calibration. Here we have to distinguish the transport model, Eq. (4.22) from the hydrodynamic model, that is compounded by Eqs. (4.17), (4.18) and (4.21), of which the main adjustment parameters are the diffusion coefficients and the source/sink terms. In the expression that parametrizes the bottom shear stress, there is a lot of uncertainty regarding the coefficient that expresses the roughness mean height, R.

In the transport model, the adjustment can be done both in the diffusion coefficients (longitudinal and transversal) and in the sources/sinks, internal (reactions) and external (contributions) [7]. For conservative substances, however, sources/sinks are not specified, which is the case for salinity, for example.

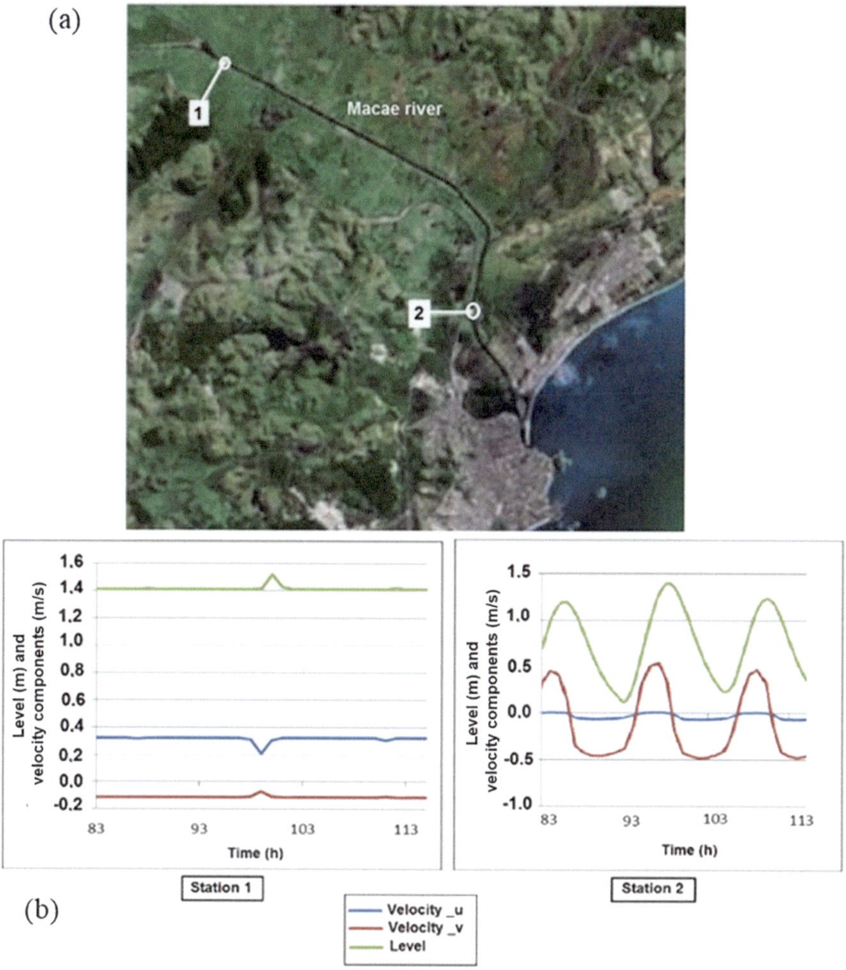

Fig. 4.4 Hydrodynamic results of monitoring stations in Macaé estuary. (**a**) Virtual monitoring stations location; (**b**) water surface level, and longitudinal (u) and transversal (v) velocity components at the virtual monitoring stations

As the fit of the models passes through confrontations with real data, the choice of the variables to be used is a critical step. In the case of the hydrodynamic model, one can choose the components of velocity or the height of the free surface, being the latter much easier to obtain in the field, and can even be done remotely. In the transport model, the variable is always the concentration of the substance to be simulated [47, 49, 71]. In this aspect, it is necessary to make a sensitivity analysis of these calculated results with respect to the parameters to be estimated. This study seeks to indicate if it is possible, as well as, what is the best data to be used in the estimation of a certain parameter, thus improving the response of the optimization

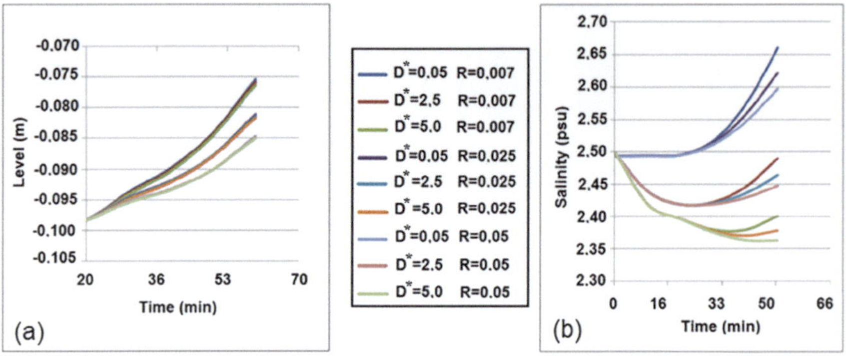

Fig. 4.5 Parametric sensitivity study for water level (**a**) and salinity (**b**)

methods employed, through the variation of the results generated by the variation of the parameters within the corresponding search spaces.

According to the classification presented in Refs. [1, 2], the roughness of the bed of rivers and estuaries subject to intense depositional regime, as in the case of the Macaé river, R can vary between 0.007 and 0.05 m. On the other hand, the noise due to the discretization (space/time) adopted, and the stability of the solutions implemented in the MOHID, imposed a range between 0.01 and 5.0 m^2s^{-1} for the turbulent diffusion coefficient D^*. The simulation results were obtained considering a period of 1 h of simulation, right after the lowest point of the tide.

Figure 4.5 synthesizes a parametric sensitivity study for the observable variables, water level and salinity, and shows that the former is not sensitive to the values adopted for the diffusion coefficient D^*. The three groups of curves are due only to the variation of the roughness height R. It can also be observed that the calculated salinity was influenced by both the diffusion coefficients and the roughness. The sensitivity evaluation was performed at the location of the virtual station 2. See Fig. 4.4a.

In Fig. 4.6a it is shown a comparison of the sensitivity coefficients, see Eq. (1.26), for the water level in relation to both parameters, D^* and R. It can be observed in Fig. 4.6b that the salinity is more sensitive to the turbulent diffusion coefficient than to the roughness mean height.

Thus, considering the observed results, it is concluded that salinity is the most indicated observable variable to be used, in order to obtain both parameters, D^* and R, simultaneously.

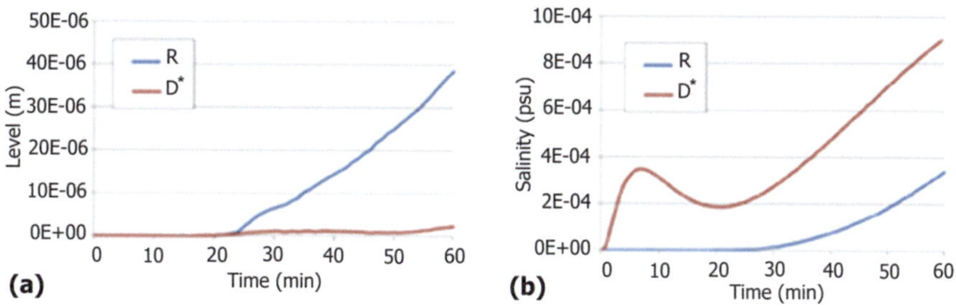

Fig. 4.6 Sensitivity coefficients (see Eq. (1.26)), for water level (**a**) and salinity (**b**) in relation to the parameters roughness and diffusion coefficient

4.3 Formulation and Solution of the Inverse Problem

4.3.1 Formulation of the Inverse Problem

The inverse problem here formulated implicitly [118, 148, 150], consists in the minimization of the objective function given by Eq. (1.23), which under some specific assumptions [15, 118] is written as

$$S(\boldsymbol{P}) = \sum_{i=1}^{N_d} \left[G_{\text{calc}_i}(\boldsymbol{P}) - G_{\text{meas}_i} \right]^2 \tag{4.24}$$

where P represents the vector of unknowns, G_{calc_i} and G_{meas_i} are the calculated and measured values, respectively, for the observable variable, with $i = 1, 2, \ldots, N_d$, and N_d is the total number of experimental data.

4.3.2 Solution of the Inverse Problem

Next are briefly presented the five methods used for the solution of the inverse problem. The first four of them, i.e. GA, PCA, LJ and PSO, are stochastic methods used for the minimization of the objective function given by Eq. (4.24). The last one is based on Artificial Neural Networks (ANN) developed for the solution of the inverse problem of interest.

Genetic Algorithms (GA)

Genetic Algorithms (GA) were developed based on the laws of natural evolution of species [68, 76, 150]. These algorithms have their philosophical basis in Darwin's survival theory of the most environmentally fit, and are based on three basic operations: selection, recombination (or crossover), and mutation. These operations act on the formation of a new generation in the execution of a GA.

The creation of the initial population is done in a random way, obeying the limits of the parameters to be estimated, i.e. within a search space previously established. The stopping criterion adopted is defined considering the achievement of a certain accuracy or the maximum number of generations. In the population evaluation step each individual is evaluated according to the objective function, Eq. (4.24), and in the next step the individuals with smaller values are more likely to be chosen as reproducers in order to generate the new population.

Particle Collision Algorithm (PCA)

This method is inspired in the interactions of a neutron in a nuclear reactor, i.e. absorption and scattering [139, 140, 150].

In the PCA a first initial configuration (solution candidate) is chosen, and then a modification of the old configuration is made to a new configuration. The two configurations are then compared. If the new solution is accepted, it serves as the old configuration for the next step. Otherwise, the algorithm continues with a new configuration change from the previous old one. Even new configurations that are worse than the previous one may be accepted in accordance with the Metropolis Criterion, similar to what happens with the Simulated Annealing method [81]. This acceptance may prevent from convergence to local minima. Since we are dealing with a minimization problem, the lower the value of the objective function, the better the configuration.

If the quality of the new configuration is better than that of the old configuration, then the "particle" is "absorbed," and a local search is performed in the neighborhood, looking for an even better solution. Otherwise, if the quality of the new configuration is worse than that of the old configuration, the "particle" may be "scattered" according to a scattering probability. By scattering we mean that the new configuration receives random values within the lower and upper bounds of each design variable. The scattering probability is inversely proportional to the quality of the solution candidate, that is, a particle with a configuration that leads to a higher value of the objective function will have a greater chance of being scattered.

Luus-Jaakola Method (LJ)

The main idea of this method is to consider at first a sufficient wide search region that encompasses the variables (feasible values), and generate random candidate solutions. Along the iterative procedure, the search region is gradually reduced around the best solution candidates [105, 150].

Particle Swarm Optimization (PSO)

The underlying idea of the algorithm is to use the experience acquired by each particle, as well as by the whole swarm, in order to determine its movement within the search space, with the computation of velocities and new positions, using [78, 150].

$$\vec{v}el_{ij+1} = \underline{w}\ \vec{v}el_{ij} + c_1 r_1 \left(\vec{p}_{best_{ij}} - \vec{p}_{ij}\right) + c_2 r_2 \left(\vec{p}_{g_j} - \vec{p}_{ij}\right) \tag{4.25a}$$

$$\vec{p}_{ij+1} = \vec{p}_{ij} + \vec{v}el_{ij+1} \tag{4.25b}$$

where $\vec{p}_{ij}$ and $\vec{v}el_{ij}$ are the position and velocity, respectively, of the particle i at the previous time step j, and, therefore $j + 1$ is the current time step, $\vec{p}_{best_{ij}}$ is the best position of the particle i up to the time step j, $\vec{p}_{g_j}$ is the best position among all particles of the swarm up to the time step j, $\underline{w}$ is the inertia coefficient, c_1 and c_2 are the cognitive coefficients that provide a balance between the individual and group experience, and r_1 and r_2 are random numbers in the range [0, 1].

Observe that in Eq. (4.25b) it is considered a unitary time step.

Artificial Neural Networks (ANN)

Artificial Neural Networks are mathematical models of computational intelligence, which have their structure based on biological neural networks, trying to extract the capacity they have to learn and generalize. This is done through the basic structures of the neural networks, the neurons. Artificial neurons can be schematized as shown in Fig. 4.7, where d_i, with i from 1 to n, are the input signals of the neuron. To these input signals weights are multiplied, representing the synapses, and these results are then summed up. The output signal is passed through a function, f_{act}, called activation function. Mathematically, the output of a neuron can be defined as

$$o = f_{act}\left(\sum_{i=1}^{n} w_i d_i\right) \tag{4.26}$$

where w_i represent the synaptic weights.

The ANN of the type Multilayer Perception (MLP) is composed by connected neurons distributed in different layers, with at least one hidden layer and the output layer, besides the entry layer in which is provided the input information [18, 67, 152]. In the supervised training, the patterns composed by known entries and the

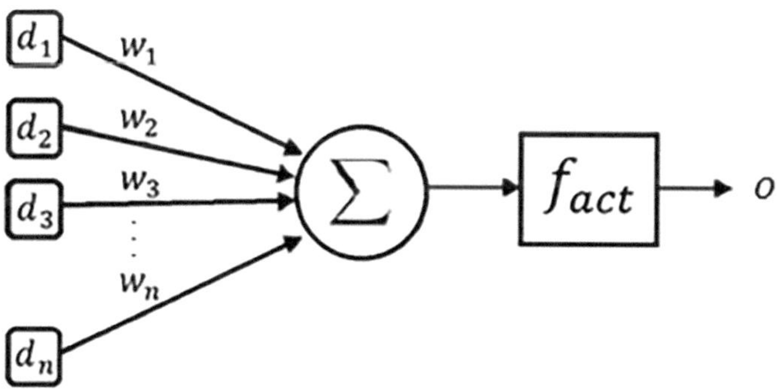

Fig. 4.7 Schematic representation of the artificial neuron

corresponding known outputs are presented to the ANN, and with an optimization algorithm the weights are computed. Afterwards, in the generalization phase, for a given set of entries, quickly the ANN provides the solution in the output layer. Such patterns may be obtained using the model considered for the direct problem of interest.

If one is interested in the solution of inverse problems, in the entry layer of the ANN is presented the experimental data, and then at the output layer the computed values correspond to the unknowns that are being estimated [150].

4.3.3 Computational Implementation

The inverse problem is here modeled with the objective of using experimental data acquired in only one single measurement station, i.e. Station 2 in Fig. 4.4a, in a relatively short period of time. The flowchart of the methodology adopted for the solution of the inverse problem formulated as an optimization problem can be visualized in Fig. 4.8.

The MOHID input data is identified with a keyword within its data files. MOHID results are written to predefined columns in the output file. So, to extract the results, it is only necessary to know in which column the desired result is. With this in hand, the sum of the quadratic error of the calculated result with respect to the experimental data is computed, being that the objective function for the optimization algorithm. The procedure is repeated until a prescribed stopping criterion is satisfied.

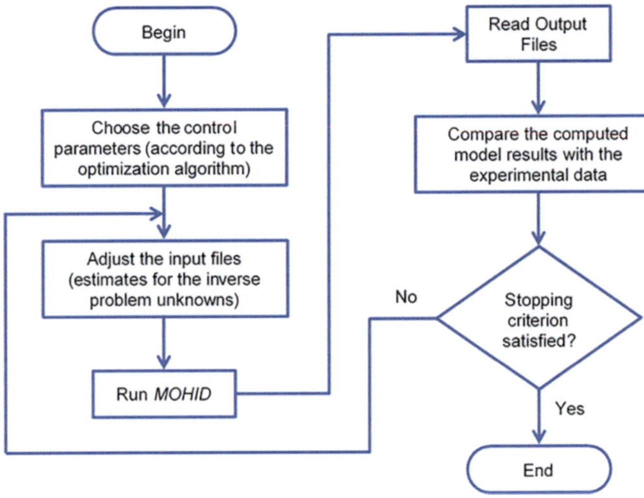

Fig. 4.8 Flowchart of the methodology for the solution of the inverse problem formulated as an optimization problem

In all cases considered in this chapter, the number of experimental data is higher than the number of unknowns. Thus, the inverse problem may be formulated as a finite-dimensional optimization problem, in which we try to minimize the norm given by the summation of the squared residues, between the model results and experimental data, as already mentioned.

In the case of the Artificial Neural Network, the patterns are generated with MOHID, and then delivered to the network training program. Once the ANN is trained, a set of experimental data is given to the input of the network, and the solution of the inverse problem is obtained.

4.4 Inverse Problem Results

The results obtained with the solution of the inverse problem employing the four different optimization methods (GA, PCA, LJ, PSO) and Artificial Neural Networks (ANN) are presented in this section. The estimation of the parameters is performed considering as experimental data, in fact, synthetic data, i.e. the computed values of the observable variable with MOHID, corrupted with Gaussian noise, adopting the values of $2.5\,\mathrm{m^2s^{-1}}$ for the coefficients of turbulent diffusion D_x^* and D_y^* in Eq. (4.22), and 0.025 m for the mean height of the bottom roughness R and the search region ranging from 0.007 to 0.05 m for this parameter, and 0.01 to $5.0\,\mathrm{m^2s^{-1}}$ for $D^* = D_x^* = D_y^*$.

The effect of the height of the bottom roughness, R, is introduced in the model by the bottom shear stress, τ_{Bx}, in Eq. (4.14) according to

$$\tau_{Bx} = \overline{U} \times \left(\frac{\kappa^*}{\ln\left(\frac{h+R}{R}\right)} \right)^2 \tag{4.27}$$

where $\overline{U}$ is the horizontal mean scalar velocity and κ^* is the von Karman constant.

4.4.1 *Genetic Algorithms (GA)*

Sixty values for the salinity were used as experimental data, being acquired at evenly distributed time intervals, considering a total period of 1 h of simulation. A total of 30 generations was used, with 5 individuals each, with a crossover probability of 50%, and 4% of mutation probability [68, 76, 150]. The error results presented in Figs. 4.9 and 4.10 correspond to the percentage error in the salinity estimation, given by

$$E_{Sl}(t) = \left| \frac{S_{l,\,\mathrm{calc}}\,(t) - S_{l,\,\mathrm{exp}}\,(t)}{S_{l,\,\mathrm{exp}}\,(t)} \right| \times 100\% \tag{4.28}$$

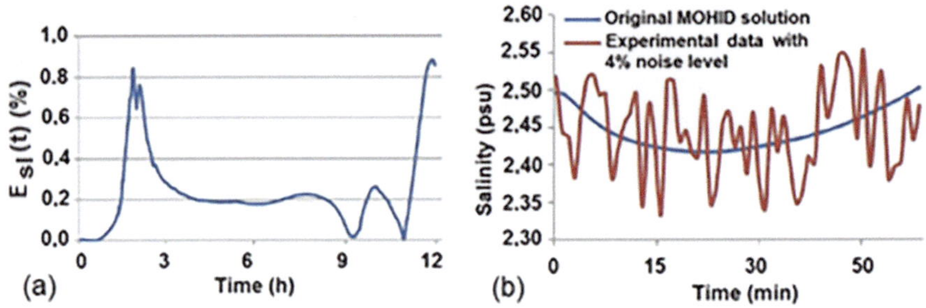

Fig. 4.9 Results obtained with GA. (**a**) Error $E_{sl}(t)$, Eq. (4.28), of the result for a 12 h time period with parameters estimated with noiseless data acquired in the first hour of simulation; (**b**) Comparison of the salinity values with and without noise for a 1 h time period of synthetic experimental data acquisition

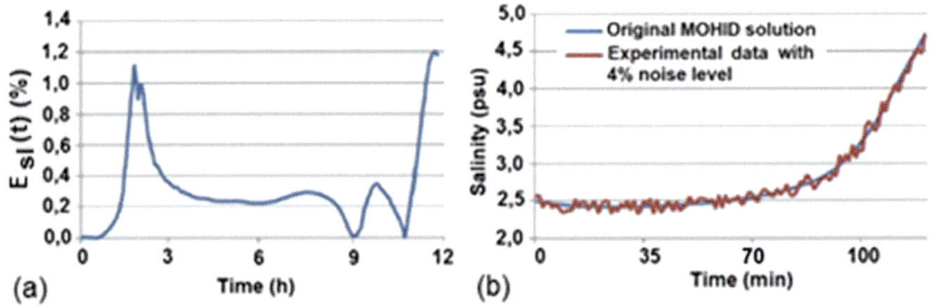

Fig. 4.10 Results obtained with GA. (**a**) Error $E_{sl}(t)$, Eq. (4.28), of the result for a 12 h time period with parameters estimated with 2 h of simulation; (**b**) Comparison of the salinity values with and without noise for a 2 h time period of synthetic experimental data acquisition

where $S_{l,calc}(t)$ and $S_{l,\exp}(t)$ corresponds, respectively, to the calculated and experimental values for the salinity at time t.

The evolution of the error with time is shown in Fig. 4.9a, for a 12-h simulation, showing that the error does not reach 1%. To evaluate the robustness of the method, a noise of 4% was added to the simulated data. Figure 4.9b shows the comparison between the original result and the data with 4% of noise level.

Due to the small amplitude of salinity variation in the period of 1 h, the added noise has a very sensitive effect on the simulated values for this observable variable. For this reason, the estimation of parameters in this example showed low efficiency, resulting in 0.0439 m for the height of the roughness and $1.7976\,\text{m}^2\text{s}^{-1}$ for the diffusion coefficient, for which the expected values were 0.025 m and $2.5\,\text{m}^2\text{s}^{-1}$, respectively, with, therefore, an estimation error above 75% for the bottom roughness height.

For this reason, the simulation period for the synthetic experimental data acquisition was increased to 2 h. See Fig. 4.10b. The error of a 12-h simulation using the estimated values obtained with the solution of the inverse problem is shown in

Fig. 4.10a. It can be observed that the error caused by the addition of noise is not significantly higher, being below 1.2%.

4.4.2 Particle Collision Algorithm (PCA)

The tests with PCA were performed considering 30 cycles for the external loop of the algorithm, and varying the number of cycles adopted in the internal loop, in order to evaluate the performance of the method [95, 140, 150].

In Fig. 4.11 are presented the results obtained (worst, best and average) for six different evaluations, being considered 5, 7 and 10 cycles in the internal loop of the PCA algorithm, and both noiseless and noisy data. The worst results were those obtained with 5 cycles. For this test with noiseless data acquired along 1 h, we obtained similarity in the average and best results, with an error lower than 0.7%. With 7 cycles we noticed a significant improvement in the best result found, whereas the results obtained for 10 cycles a slight degradation in the results was observed.

As in the cases performed with GA, a 4% noise level in the experimental data was used to test the method, and as in those tests, the data was taken in a 2 h timeframe of simulation starting at the lowest tide. The result for 5 cycles was very poor for the worst case. It is also observed that, although the worst result has a very large error, the average result showed an error below 2%. Differently from the results found when noiseless data was used, the test response using 7 cycles presented worse results for the worst case than those obtained with 5 cycles. For this case, when tested with 10 cycles, the results improved significantly, with errors below 1.2% in the best and the average results.

4.4.3 Luus-Jaakola Method (LJ)

In order to maintain the same number of objective function evaluations adopted for the other methods, it was decided to use $n_{out} = 30$ and $n_{int} = 5$ for the number of cycles in the external and internal loops, respectively, in the Luus-Jaakola method [105, 150]. The contraction coefficient used was 0.05 and, as in the other previous cases considered, 10 tests were performed for the estimation of the parameters, in order to verify the stability of the responses of the method.

In Fig. 4.12a are shown the errors in a 12 h period of simulation, being the parameters estimated with the LJ method, using noiseless data acquired at the first 1 h timeframe. The results obtained with 4% noise level experimental data acquired at the first 2 h timeframe are shown in Fig. 4.12b. In the former it can be observed that the error remained below 2%, while in the latter it reached a maximum value close to 2.5%.

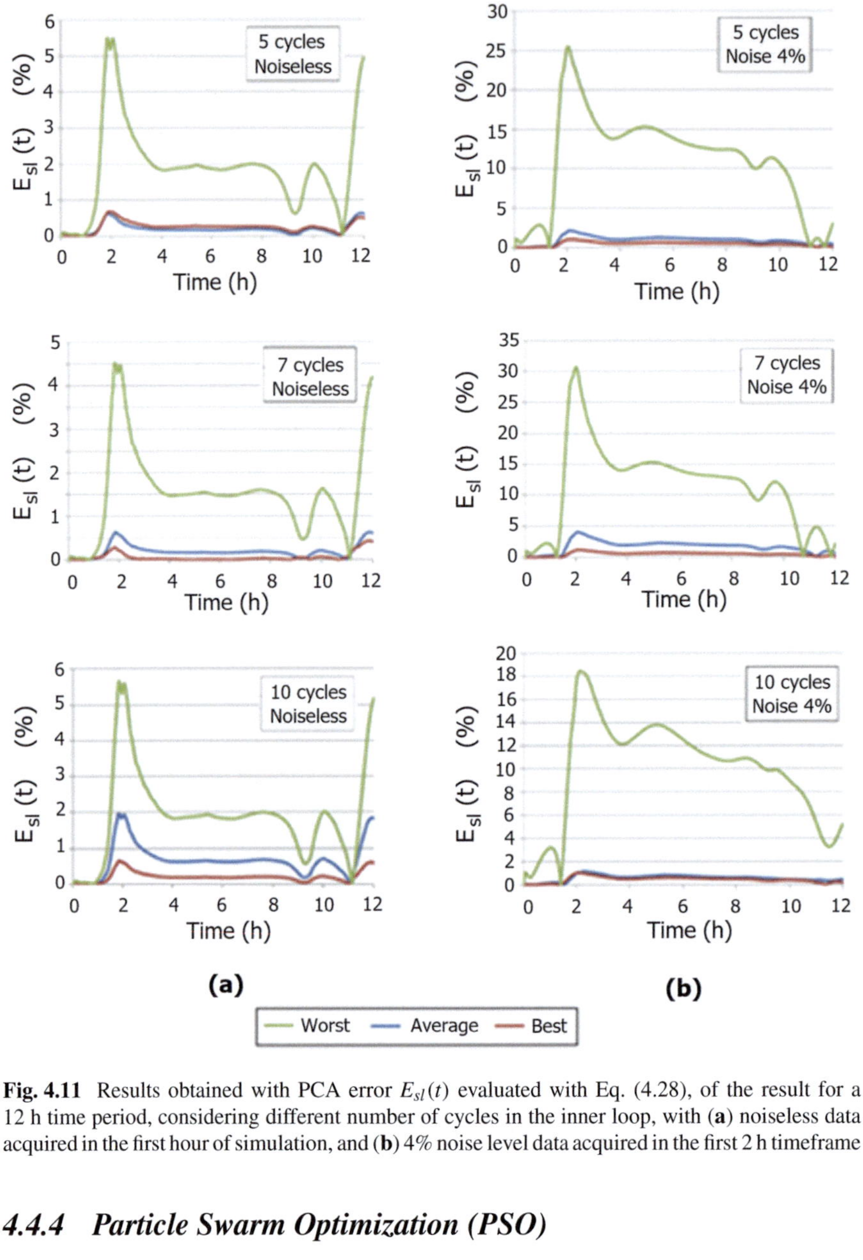

Fig. 4.11 Results obtained with PCA error $E_{sl}(t)$ evaluated with Eq. (4.28), of the result for a 12 h time period, considering different number of cycles in the inner loop, with (**a**) noiseless data acquired in the first hour of simulation, and (**b**) 4% noise level data acquired in the first 2 h timeframe

4.4.4 Particle Swarm Optimization (PSO)

For the test cases with PSO, 10 particles were adopted, and a total of 15 particle movements, i.e. 15 time steps, which differs from the 5/30 particles/individuals and/ or cycles configuration adopted for the methods GA, PCA and LJ. For the inertia

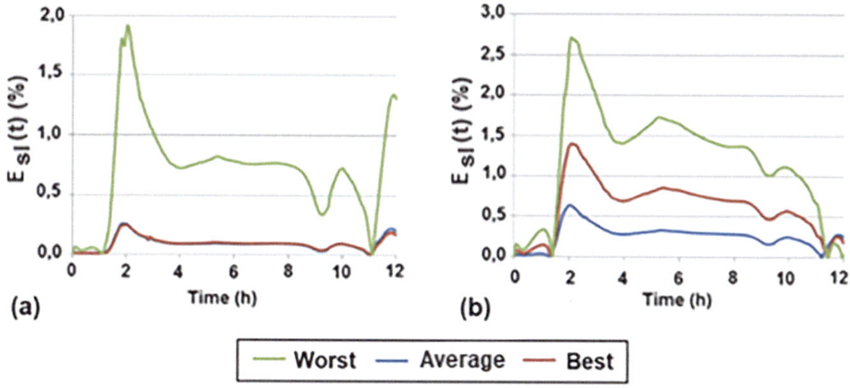

Fig. 4.12 Results obtained with the LJ method, Error $E_{sl}(t)$, evaluated with Eq. (4.28), of the result for a 12 h time period, with parameters estimated with (**a**) noiseless data acquired in the first hour of simulation, and (**b**) 4% noise level data acquired in the first 2 h timeframe

constant it was used $\underline{w} = 0.61$, and for the cognitive coefficients the following values were adopted $c_1 = c_2 = 2$ [78, 150].

The results obtained with such configuration, using noiseless data for the estimation of the unknown parameters, presented a good performance, reaching the exact value in some cases. In spite of that, as can be seen in Fig. 4.13a for the worst result the errors reached up to 6%, a value higher than that found in the worst case with the PCA, see Fig. 4.11a. In contrast, when data with a noise level of 4% was used, see Fig. 4.13b, the error for the worst case remained below 6%, and the other cases, i.e. best and average, remained with errors below 1.2%. These results show that the PSO has the ability to reach its goal even with a small number of iterations.

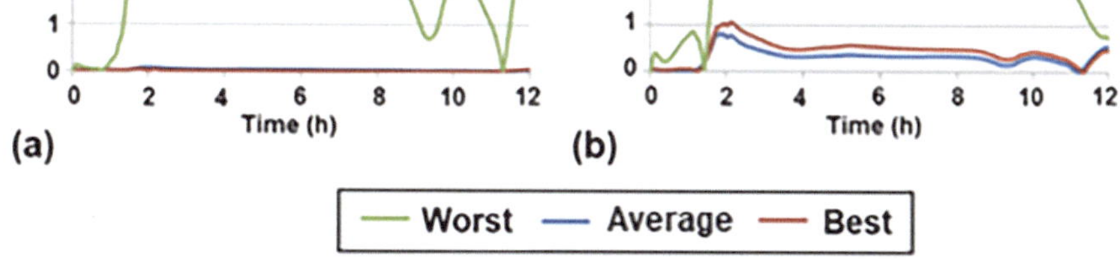

Fig. 4.13 Results obtained with the PSO algorithm. Error $E_{sl}(t)$, evaluated with Eq. (4.28), of the result for a 12 h time period with parameters estimated with (**a**) noiseless data acquired in the first hour of simulation, and (**b**) 4% noise level data acquired in the first 2 h timeframe

Table 4.1 Exact values considered for the height of the bottom roughness *R*, and for the turbulent dispersion coefficient D^*

Test case	R (m)	$D^*(m^2s^{-1})$
1	0.007	0.1
2	0.007	5.0
3	0.050	0.1
4	0.050	5.0
5	0.025	2.5

4.4.5 *Artificial Neural Networks (ANN)*

The input data for the ANN [67, 150] consists of the salinity variation in the first 2 h timeframe. The network was trained with patterns generated as follows: (a) 10 values for R (varying from 0.007 to 0.05 m), and 10 values for D^* (ranging from 0.01 to 5.0 m^2s^{-1}) yielding, therefore, 100 patterns; (b) for each pattern five new patterns were generated, with the addition of artificial noise corresponding to 1, 2, 3, 4 and 5% noise level, totaling 500 new training patterns. With this, the ANN was defined with 60 input neurons, represented by 60 salinity values, measured at a station during a period of 2 h starting at the time the lowest tide occurred; and 2 output neurons, represented by the mean height of the roughness R and the turbulent dispersion coefficient D^*. The value of the output neurons was normalized to have their answers in the range from -1 to 1.

In Table 4.1 are shown the values used for R and D^* in five different test cases.

The results obtained with the ANN for different input data (noiseless and 4% noise level data), and different ANN configurations (3 and 5 layers), are presented in Tables 4.2 and 4.3.

In Table 4.2, are shown the results for the five test cases considering a 3-layers ANN for noiseless data, and the error in obtaining D^* reaches up to 69.5%. For this reason, the maximum error in a 12-h simulation using the values estimated for the parameters with the solution of the inverse problem can reach close to 10% in the first case and 3% in the third one. The lower error in the third case is due to the fact that, although the diffusion coefficient was poorly estimated, the good roughness mean height R estimation was able to guarantee a good simulation generated by the hydrodynamic model, when adopted a longer time interval. When 4% noise level data is considered, good estimates are obtained for the unknowns, yielding reasonably low values for the salinity along a 12 h time period of observation.

In Table 4.3 are shown the results obtained with a 5-layers ANN. It is possible to observe an overall improvement on the inverse problem solution.

Table 4.2 Results for the estimation of the bottom roughness mean height R, and of the turbulent dispersion coefficient D^*, with a 3-layers ANN, considering noiseless and 4% noise level data acquired in the first 2 h timeframe

Test cases	Noiseless data	4% Noise level data
1	$\frac{\hat{R}}{R} = 1.2571$ $\frac{\hat{D}^*}{D^*} = 1.6950$ $E_{max} = 9.71\%$	$\frac{\hat{R}}{R} = 1.1857$ $\frac{\hat{D}^*}{D^*} = 1.5780$ $E_{max} = 7.40\%$
2	$\frac{\hat{R}}{R} = 1.0429$ $\frac{\hat{D}^*}{D^*} = 0.9730$ $E_{max} = 1.00\%$	$\frac{\hat{R}}{R} = 1.1857$ $\frac{\hat{D}^*}{D^*} = 0.9853$ $E_{max} = 3.01\%$
3	$\frac{\hat{R}}{R} = 0.9460$ $\frac{\hat{D}^*}{D^*} = 1.6870$ $E_{max} = 2.84\%$	$\frac{\hat{R}}{R} = 0.9180$ $\frac{\hat{D}^*}{D^*} = 1.2720$ $E_{max} = 1.95\%$
4	$\frac{\hat{R}}{R} = 0.9740$ $\frac{\hat{D}^*}{D^*} = 1.6870$ $E_{max} = 0.68\%$	$\frac{\hat{R}}{R} = 0.9240$ $\frac{\hat{D}^*}{D^*} = 0.9472$ $E_{max} = 1.82\%$
5	$\frac{\hat{R}}{R} = 0.9960$ $\frac{\hat{D}^*}{D^*} = 0.9956$ $E_{max} = 0.15\%$	$\frac{\hat{R}}{R} = 0.9760$ $\frac{\hat{D}^*}{D^*} = 1.1488$ $E_{max} = 3.67\%$

$\hat{R}$, $\hat{D}^*$—estimated values; R, D^*—exact values of the unknown parameters; E_{max}—value of the maximum error on the salinity $E_{sl}(t)$, obtained with Eq. (4.28)

4.5 Remarks

The strategies developed in this chapter have presented great potential as a tool for the calibration of hydrodynamic models. With regard to the applied optimization methods, it can be verified that:

- the GA approach proved to be quite efficient for estimating the parameters of roughness height and turbulent diffusion coefficients, resulting in an error calibration of less than 1.2%.
- the results obtained with the PCA were not conclusive as to the effectiveness in increasing the number of cycles in the inner loop. Despite that, good estimates for the unknowns were obtained in 80% of the cases, showing that this approach, although it may present very large errors in some runs, has the great possibility of generating good results.
- the LJ method, although the simplest one in relation to the others considered in this chapter, presented the best overall stability of the results, with different runs of the method, yielding an error of less than 3%, even in the worst case.

Table 4.3 Results for the estimation of the bottom roughness mean height R, and of the turbulent dispersion coefficient D^*, with a 5-layers ANN, considering noiseless and 4% noise level data acquired in the first 2 h timeframe

Test case	Noiseless data	4% Noise level data
1	$\frac{\hat{R}}{R} = 1.2143$ $\frac{\hat{D}^*}{D^*} = 1.4803$ $E_{\max} = 8.17\%$	$\frac{\hat{R}}{R} = 1.0428$ $\frac{\hat{D}^*}{D^*} = 1.6700$ $E_{\max} = 4.34\%$
2	$\frac{\hat{R}}{R} = 1.0714$ $\frac{\hat{D}^*}{D^*} = 0.9820$ $E_{\max} = 1.42\%$	$\frac{\hat{R}}{R} = 1.0571$ $\frac{\hat{D}^*}{D^*} = 0.9843$ $E_{\max} = 1.08\%$
3	$\frac{\hat{R}}{R} = 0.9320$ $\frac{\hat{D}^*}{D^*} = 1.5240$ $E_{\max} = 2.27\%$	$\frac{\hat{R}}{R} = 0.9420$ $\frac{\hat{D}^*}{D^*} = 1.1350$ $E_{\max} = 1.25\%$
4	$\frac{\hat{R}}{R} = 0.9800$ $\frac{\hat{D}^*}{D^*} = 0.9533$ $E_{\max} = 0.97\%$	$\frac{\hat{R}}{R} = 0.9700$ $\frac{\hat{D}^*}{D^*} = 0.9161$ $E_{\max} = 1.78\%$
5	$\frac{\hat{R}}{R} = 1.0201$ $\frac{\hat{D}^*}{D^*} = 1.0026$ $E_{\max} = 0.51\%$	$\frac{\hat{R}}{R} = 1.0200$ $\frac{\hat{D}^*}{D^*} = 0.9993$ $E_{\max} = 0.48\%$

$\hat{R}$, $\hat{D}^*$—estimated values; R, D^*—exact values of the unknown parameters; $E_{\max}$—value of the maximum error on the salinity $E_{sl}(t)$, obtained with Eq. (4.28)

- the PSO presented very good, even almost exact, results in many of its responses when the parameters were estimated with noiseless data. This excellent performance was not verified in the tests with noisy data, presenting results that were equal, or worse than the ones obtained with the PCA method.
- the ANN proved to be a good tool to estimate the parameters of a model that simulates drainage and transport in estuaries when properly trained. Even when the estimates for the unknown parameters were relatively poor, the results for the salinity represented a good approximation, with not so large computed errors for the 12 h time period of observation. The use of noisy data did not produce an effectively worse result, even using noise levels different from those used in the training phase of the ANN. In addition, for the 5 layers configuration, the use of noisy data led to even better results when compared to noiseless data. This shows that a 5-layers ANN configuration has a higher generalization power with noisy data for the test cases performed.

In estuarine models the river parameters differ from the rest of the estuary parameters, and although only one parameter value was used for the whole estuary region in the chapter, it is possible to identify with the methods presented, a larger number of parameters, with a larger number of measuring stations, properly distributed throughout the domain of interest.

Chapter 5
Estimation of Pollutant Sources Intensity and Location in Estuarine Environments

In this chapter Inverse Problem techniques are applied to estimate the source parameters (location and intensity) of a hypothetical conservative pollutant released in estuarine waters. The case study considered here is the Macaé River estuary, located in the Brazilian southeast coast. The pollutant transport was modeled by the advection–diffusion equation. For the source location estimation was used a Computational Intelligence approach, with the application of the Luus-Jaakola (LJ), Particle Collision Algorithm (PCA) and Ant Colony Optimization (ACO) methods, while for the estimation of the source intensity was used the Golden Section method (GSM).

5.1 Introduction

The case study presented here is the Macaé River estuary, located on the northern coast of the state of Rio de Janeiro. The preservation of such environments is justified by the great biological diversity, due to the hydrodynamic characteristics of the predominant circulation, which, imprisoning nutrients, algae and other plants, stimulates the productivity of these water bodies [114].

The proposed problem consists in the identification of the origin and magnitude of a hypothetical pollutant release that is diluted in the waters of this estuary. For that purpose, the solution of the direct problem, modeled by the transport equation, is implemented. Such model describes the behavior of a contaminant, involving hydrodynamic parameters, dispersion and a term representing sources or sinks. The direct problem model is coupled to Computational Intelligence methods for the estimation of parameters (inverse problem) that can express the location and intensity of

---

*Radael de Souza Parolin, from UNIPAMPA, Pampa Federal University, co-authored this chapter.

J. Lugon Junior et al., *Inverse Problems in the Environmental Modelling of Rivers and Estuaries*, Studies in Systems, Decision and Control 246,
https://doi.org/10.1007/978-3-032-09937-2_5

the source of the pollutant considered. Therefore, the main objective is to identify pollutant sources through such parameters.

5.2 Formulation and Solution of the Direct Problem

The transport of constituents can be described by the advection–diffusion equation, Eq. (1.20), which can be derived from the constituent mass conservation equation, which states that the rate of change for a scalar quantity in a differential control volume is given by the flux through the boundaries along with diffusion and generation or consumption within the control volume [175]. The mathematical formulation of such problem may be written as

$$\frac{\partial C}{\partial t} + \nabla\left(\vec{V} C\right) + S_c - R_c = 0 \tag{5.1}$$

where C is the concentration of the constituent, $\vec{V}$ is the total mass flux, and S_c and R_c are the source and sink terms, respectively.

The total mass flux is given by

$$\vec{V} = \vec{v} C - D^{*}\nabla C \tag{5.2}$$

where $\vec{v}$ is the fluid velocity vector with the components u, v and w in the x, y and z directions, respectively, and D^{*} is the turbulent diffusion coefficient.

Considering that estuaries have shallow water, and that the concentration vary little in the vertical coordinate, it is then approximated as vertically homogeneous. Thus, to eliminate the vertical coordinate, the following integration is applied

$$\int_h \left(\frac{\partial C}{\partial t} + \vec{v}\nabla C - \nabla\left(D^{*}\nabla C\right) + S_c - R_c\right) dz = 0 \tag{5.3}$$

Assuming that the concentration C and the horizontal velocity components u and v are no longer dependent on the vertical spatial coordinate z, and that the vertical component of velocity w is zero, results,

$$h\frac{\partial C}{\partial t} + h\vec{v}\nabla C - \nabla\left(hD^{*}\nabla C\right) + h(S_c - R_c) = 0 \tag{5.4}$$

where h is the total depth of the water column.

The boundary conditions (Dirichlet and Neumann conditions) and initial condition are given as

$$C(x_D, y_D, t) = C_D(x_D, y_D, t) \tag{5.5}$$

$$D^* \frac{\partial C(x_N, y_N, t)}{\partial \vec{n}} = f_N(x_N, y_N, t) \tag{5.6}$$

$$C(x, y, 0) = C_{ini}(x, y) \tag{5.7}$$

where (x_D, y_D) and (x_N, y_N) are points located at the boundaries of the domain in which Dirichlet and Neumann boundary conditions apply, respectively, D^* is the matrix with turbulent diffusion coefficients, $\vec{n}$ is the unit outward normal vector at the boundaries, C_D are the known concentrations for the Dirichlet boundary condition, f_N are the known mass fluxes for Neumann's boundary condition, and C_{ini} are the known concentration values for the initial condition.

5.2.1 Numerical Solution

The discretization of the transport model is here performed with the application of the Finite Element Method (FEM) in the spatial coordinate [43, 198], and the Finite Difference Method (FDM) in time [131, 175]. The application of the FEM transforms a partial differential equation (PDE) into a system of ordinary differential equations (ODEs) and the application of the FDM transforms each ODE system into a system of algebraic linear equations, written in terms of vectors and matrices.

The FEM is a particular case of the Weighted Residual Method (WRM) [72], in which the following integral equation must be satisfied

$$\iint_{\Omega} \left[h\frac{\partial C}{\partial t} + h\vec{v}\,\nabla C - \nabla\big(hD^*\nabla C\big) + h(S_c - R_c) \right] M_i(x, y) d\Omega = 0 \tag{5.8}$$

$$i = 1, 2, \ldots, N_n$$

where $M_i(x, y)$ are the weighting functions associated to the node i of a finite element mesh, with a total of N_n nodes, that represents the study region Ω.

Decomposing the expression above, and applying the two-dimensional integration by parts, one can write in a restricted form, to a finite element e, the following vector equation [72]:

$$\sum_{e=1}^{N_e} \left(I_{1i}^e + I_{2i}^e + I_{3i}^e + I_{4i}^e\right) - \sum_{e=1}^{N_{bs}} \left(I_{5i}^e\right) = 0, i = 1, 2, \ldots, N_e \tag{5.9}$$

where N_e is the total number of elements, N_{bs} is the total number of boundary surfaces of the elements, and the components of the vectors of each term are given by the following integrals

$$I_{1i}^{e} = \iint_{\Omega_e} h\frac{\partial C}{\partial t} M_i d\Omega_e \tag{5.10a}$$

$$I_{2i}^{e} = \iint_{\Omega_e} h\vec{v}\,\Delta C M_i d\Omega_e \tag{5.10b}$$

$$I_{3i}^{e} = -\iint_{\Omega_e} (hD^{*}\nabla C)\nabla M_{i} d\Omega_e \tag{5.10c}$$

$$I_{4i}^{e} = \iint_{\Omega_e} h(S_c - R_c) M_i d\Omega_e \tag{5.10d}$$

$$I_{5i}^{e} = \oint_{\partial\Omega_e} (hD^{*}\nabla C\vec{n}) M_i d\Omega_e, \tag{5.10e}$$

with $i = 1, 2, \ldots, N_e$, where Ω_e represents each element physical domain, $\partial\Omega_e$ represents the boundary of the finite element e, and $\vec{n}$ is the unit outward normal vector at the boundary.

In order to calculate the integrals, it is necessary to approximate the function within the finite element, and considering a linear interpolation results

$$C_e(x, y, t) \approx \sum_{j=1}^{N_n^e} C_j(t) N_j(x, y) \tag{5.11}$$

where $N_j(x, y)$ are the base functions associated to node j of the mesh; $C_j(t)$ is the concentration time dependence in node j, and N_n^e is the number of nodes in element e.

The derivatives with respect to the spatial and temporal coordinates are

$$\frac{\partial C_e(x, y, t)}{\partial x} = \sum_{j=1}^{N_n^e} C_j(t) \frac{\partial N_j(x, y)}{\partial x} \tag{5.12a}$$

$$\frac{\partial C_e(x, y, t)}{\partial y} = \sum_{j=1}^{N_n^e} C_j(t) \frac{\partial N_j(x, y)}{\partial y} \tag{5.12b}$$

$$\frac{\partial C_e(x, y, t)}{\partial t} = \sum_{j=1}^{N_n^e} \frac{\partial C_j(t)}{\partial t} N_j(x, y) \tag{5.12c}$$

Therefore,

$$\nabla C_e(x, y, t) = \sum_{j=1}^{N_n^e} C_j(t) \nabla N_j(x, y) \tag{5.12d}$$

It is observed in Eq. (5.11) that the function representing the concentration $C_e(x, y, t)$, is decomposed into the product of the concentration time dependence at each node by the base functions associated to each node. In this way, the variables are separated, since the base functions depend only on the spatial variables, and the

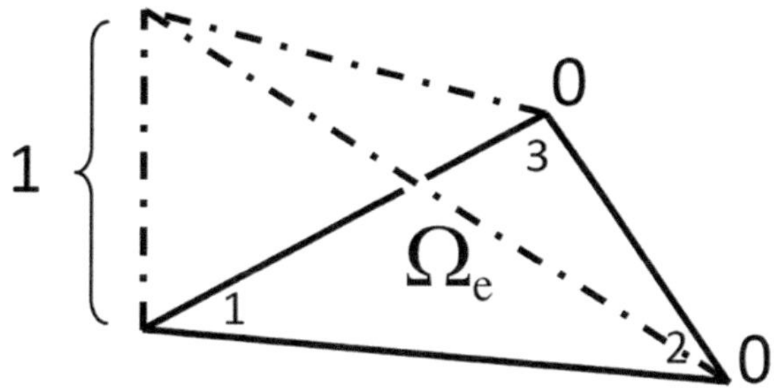

Fig. 5.1 Triangular finite element with the linear interpolation conditions of the base function $N_1(x, y)$

concentration values depend only on the time variable. Consequently, the derivatives represented in Eqs. (5.12a-d) have the same characteristic of separating the spatial and temporal terms.

Here the finite element is considered as a triangle, and the base function associated with node j is defined as a linear polynomial,

$$N_j(x, y) = \overline{a}_j x + \overline{b}_j y + \overline{c}_j \tag{5.13}$$

in which the coefficients $\overline{a}_j$, $\overline{b}_j$ and $\overline{c}_j$ depend on the coordinates of the three vertices of the triangle, and are calculated using the interpolation conditions. For example, the base function associated with node $j = 1$, must satisfy the following conditions, as shown in Fig. 5.1.

$$N_1(x_1, y_1) = \overline{a}_1 x_1 + \overline{b}_1 y_1 + \overline{c}_1 = 1 \tag{5.14a}$$

$$N_1(x_2, y_2) = \overline{a}_1 x_2 + \overline{b}_1 y_2 + \overline{c}_1 = 0 \tag{5.14b}$$

$$N_1(x_3, y_3) = \overline{a}_1 x_3 + \overline{b}_1 y_3 + \overline{c}_1 = 0 \tag{5.14c}$$

The weight functions, $M_i(x, y)$, and the base functions, $N_j(x, y)$, can be defined in different ways. Considering the relation of Bubnov-Galerkin (BG), or referred to as the relation of Galerkin in FEM, is given by [72]:

$$M_j(x, y) = N_j(x, y) \tag{5.15}$$

The Petrov–Galerkin (PG) relation is given by

$$M_j(x, y) = N_j(x, y) + \frac{\beta h^e}{2\left|\overrightarrow{v}\right|}\left(u\frac{\partial N_j(x, y)}{\partial x} + v\frac{\partial N_j(x, y)}{\partial y}\right) \tag{5.16}$$

where h^e is a characteristic length of the element e, $\left|\overrightarrow{v}\right|$ is the modulus of the velocity vector, and β is a constant given by

$$\beta = coth\,coth(Pe) - \frac{1}{Pe} \tag{5.17}$$

$$Pe = \frac{|\overrightarrow{v}|h^e}{\sqrt{det(D^*)}} \tag{5.18}$$

where Pe is the Peclet number [91, 198], which consists on the relation between advection and diffusion, the first one being represented by the speed $\overrightarrow{v}$ and the characteristic length h^e, and the second one by the dispersion coefficients D^*.

With the BG formulation, the method is conditionally stable for $Pe< 1$, and with the PG formulation, the method is unconditionally stable for any Pe number [198]. It is observed that the PG relation is equal to the BG relation in the case of $\beta = 0$.

5.2.2 Mesh Analysis

In order to define a mesh to be used it is necessary to analyze consistency, stability and convergence issues of the discretization method used. With the discretization of the Partial Differential Equation (PDE) of advection–diffusion, Eq. (5.4), and the computers finite arithmetic, errors are generated. These are the discretization errors (DE) and the rounding errors (RE), respectively.

Discretization error is the difference between the PDE exact analytical solution and the exact discrete solution (EDS), with no rounding errors. The discretization error is simply the truncation error (TE) plus errors introduced by the boundary conditions numerical treatment [175]. Thus, ignoring such errors by numerical treatment, TE = PDE–EDS.

Rounding error is the numerical error introduced by the high number of calculations in which the computer constantly rounds numbers due to its finite arithmetic [175]. Thus, the discretized numerical solution (DNS) has rounding errors, ie, RE = EDS–DNS. In some types of computations, the magnitude of these errors is proportional to the number of mesh points in the problem domain. In these cases, refining the mesh may decrease TE, but increase RE [175].

In general, due to the finite arithmetic of computers, we are always dealing with DE and RE at the same time. Thus, PDE–DNS = DE + RE.

The consistency is related to the proximity of the exact discretized solution (EDS) with the exact PDE solution, without considering rounding errors. Thus, the discretized representation of a PDE is consistent if it can be shown that the difference between the PDE and its discretized representation vanishes with the refinement of the mesh [175], that is,

$$(PDE - DNS) = TE = 0 \tag{5.19}$$

Numerical stability is a concept applied only to marching problems. In a stable numerical scheme, errors of any source (rounding, truncation, numerical processing) do not grow in numerical procedures from one temporal iteration to another [174]. To prove the stability and behavior of the temporal integration algorithm, a very convenient procedure is to compare exact (analytical) and numerical solutions. The convergence occurs when the discrete solution approaches the exact PDE solution (with the same boundary and initial conditions) as the mesh is refined. For problems of initial value, consistency and stability imply convergence, according to Lax's equivalence theorem [175].

For the application of the FEM using the PG and BG there are stability conditions that must be satisfied [160, 164].

In the search for a consistent and convergent numerical discretization, besides being feasible in terms of computational cost, different simplified meshes were proposed, and the calculated concentrations were compared with an analytical solution. To verify stability, the W-weight parameter of the FDM was optimized for each proposed mesh by formulating an inverse problem. Considering a particular case without sources and sinks, and advection only in the x direction, from Eqs. (5.1) and (5.2), see also Eq. (1.20), the transport problem is given by [91]

$$\frac{\partial C}{\partial t} + u\frac{\partial C}{\partial x} - \frac{\partial}{\partial x}\left(D_{xx}\frac{\partial C}{\partial x} + D_{xy}\frac{\partial C}{\partial y}\right) - \frac{\partial}{\partial y}\left(D_{yx}\frac{\partial C}{\partial x} + D_{yy}\frac{\partial C}{\partial y}\right) = 0 \quad (5.20)$$

The proposed analytical manufactured solution [88] for the two-dimensional domain shown in Fig. 5.2, with length $L_x = 10{,}000$ m and $L_y = 4{,}000$ m is given by.

$$C_{ams}(x, y, t) = C_b - \frac{2D_{xx}}{\beta_1^2} - \frac{2D_{yy}}{\beta_2^2} - \frac{(x-ut)^2}{\beta_1^2} - \frac{1}{\beta_2^2}\left(y - \frac{L_y}{2}\right)^2 \quad (5.21)$$

where $C_b = 10\,\text{kgm}^{-3}$, $\beta_1 = 5{,}000\,\text{m}^{\frac{5}{2}}\text{kg}^{-\frac{1}{2}}s^{-\frac{1}{2}}$, and $\beta_2 = 900\,\text{m}^{\frac{5}{2}}\text{kg}^{-\frac{1}{2}}s^{-\frac{1}{2}}$.

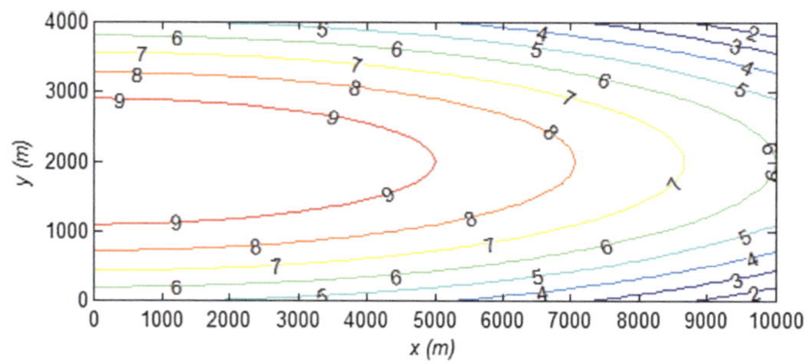

Fig. 5.2 Initial condition concentrations ($kg\,m^{-3}$)

A feature of this solution is that it does not tend to a stationary solution when $t \rightarrow \infty$ (concentrations become negative). However, the analytical and calculated data are compared for specific times. For the calculated problem we have the following first type, Dirichlet boundary conditions and initial condition,

$$C(0, y, t) = C_{ams}(0, y, t), \text{ for } 0 \leq y \leq 4{,}000 \text{ m and } t > 0 \tag{5.22a}$$

$$C(10{,}000, y, t) = C_{ams}(1{,}000, y, t), \text{ for } 0 \leq y \leq 4{,}000 \text{ m, and } t > 0 \tag{5.22b}$$

$$C(x, 0, t) = C_{ams}(x, 0, t), \text{ for } 0 \leq x \leq 10{,}000 \text{ m, and } t > 0 \tag{5.22c}$$

$$C(x, 4{,}000, t) = C_{ams}(x, 4{,}000, t), \text{ for } 0 \leq x \leq 10{,}000 \text{ m, and } t > 0 \tag{5.22d}$$

$$C(x, y, 0) = C_{ams}(x, y, 0), \text{ for } 0 \leq x \leq 10{,}000 \text{ m, and } 0 \leq y \leq 4{,}000 \text{ m} \tag{5.22e}$$

The physical parameters considered in this case are $v = 1 \text{ m s}^{-1}$, $v = 0 \text{ m s}^{-1}$, $D_{xx} = D_{yy} = 2 \text{ m}^2 \text{ s}^{-1}$, and $D_{xy} = D_{yx} = 0 \text{ m}^2 \text{ s}^{-1}$. In Fig. 5.2 the initial condition of the proposed problem ($t = 0s$) is presented.

The proposed spatial meshes are uniform, composed of isosceles triangles, constructed using $\Delta x = \Delta y$. For the analysis of this mesh, three spatial increments were considered, i.e. $\Delta x = \Delta y = 100$ m, 500 m, 1,000 m. In Fig. 5.3 the mesh with $\Delta x = \Delta y = 500$ m is shown.

In Fig. 5.4, the concentrations at $y = 2{,}000$ m are presented for three instants of time, $t = 3{,}600$ s, 7,200 s and 10,800 s, with the use of the analytical solution and the calculated solution for the three different proposed spatial meshes.

For the analysis of the temporal mesh, five temporal increments were considered, $\Delta t = 300$ s, 600 s, 900 s, 1,800 s and 3,600 s, with a fixed spatial discretization, using *Δx = Δy = 500 m*. In Fig. 5.5, the concentrations at $y = 2{,}000$ m are presented for three time instants with the use of the analytical solution and the calculated

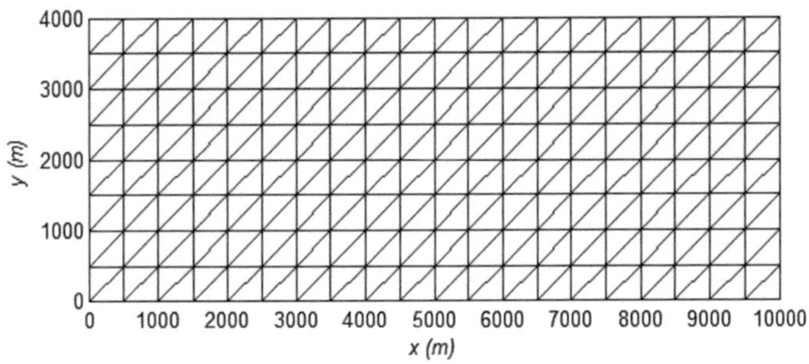

Fig. 5.3 Uniform spatial mesh with ($\Delta x = \Delta y = 500$ m)

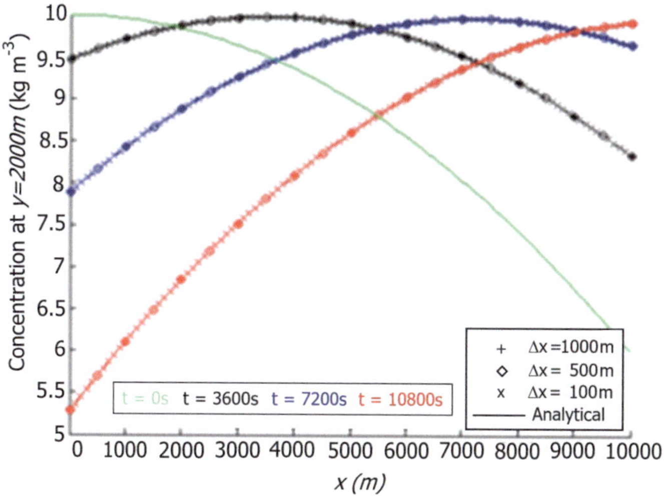

Fig. 5.4 Concentrations at y = 2,000 m for different spatial meshes

solution for the five different proposed temporal meshes. The number of time steps, N_t, was taken for each temporal mesh in such way that the final time was the same for all cases, i.e. 10,800 s.

In Table 5.1 are shown the values of the root mean square error (RMSE) between the calculated and analytical concentrations. The RMSE has been defined in Eq. (3.44), in the context of inverse problems. As can be seen, the error is smaller

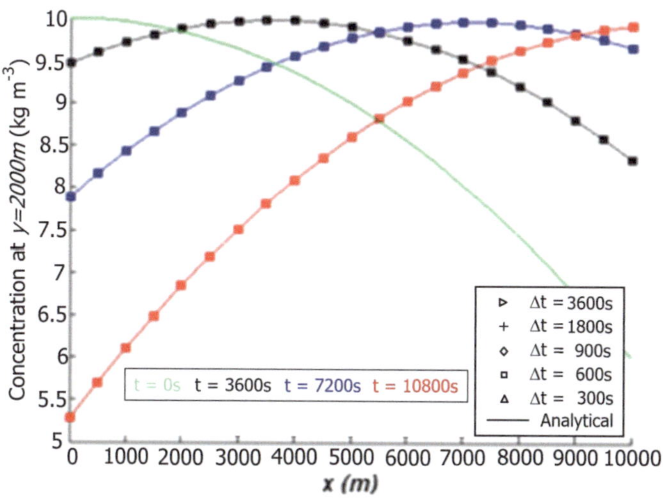

Fig. 5.5 Concentrations at y = 2,000 m for different temporal meshes

Table 5.1 RMSE for different spatial and temporal meshes

Spatial mesh	$\Delta x = \Delta y$			Temporal mesh	Δt				
	1,000 m	500 m	100 m		3,600 s	1,800 s	900 s	600 s	300 s
N_e N_n	80 55	320 189	8,000 4,141	N_t	3	6	12	18	36
RMSE $[kgm^{-3}]$	$< 10^{-4}$	$< 10^{-4}$	$< 10^{-4}$	RMSE $[kgm^{-3}]$	$< 10^{-4}$	$< 10^{-4}$	$< 10^{-4}$	$< 10^{-4}$	$< 10^{-4}$

N_e–Number of elements, N_n–Number of nodes, N_t–Total number of time steps considered

than 10^{-4} kg m^{-3} using the three spatial meshes and the five temporal meshes. That is corroborated by the superposition of the results shown in Figs. 5.4 and 5.5.

Consistency, stability and, therefore, convergence are demonstrated for the test case considered, with the analytical manufactured solution.

5.2.3 *Macaé Estuary Geometry and Bathymetry*

The domain of study, which corresponds to the Macaé River estuary [181], located in the municipality of Macaé in the State of Rio de Janeiro, was defined by satellite photos, extending for almost 20 km, from the headwaters to the coast. The bathymetric data (depth according to mean sea level) of the coastal region were taken from the nautical chart 1507, edited by the Brazilian Navy in 1974 [22], while data for the upper region was obtained from Ref. [7]. Due to the existence of few available bathymetric data points from the upper region, linearly interpolated data were used. For the coast, the data were referenced with the closest existing bathymetry information. It should be noted also that these data are out of date due to the various changes in the estuary, and in the vicinity, caused by human action in recent decades. The geometry of the estuary and the bathymetry data used are shown in Fig. 5.6. The space grid is composed of 600 nodes (281 belonging to the boundary) with 917 finite triangular elements.

5.2.4 *Parameters of the Transport Model and Boundary Conditions*

The transport model contains dispersion coefficients, which in practice can vary greatly, both in experimental results in natural environments and in theoretical estimates [45]. The values considered for these coefficients were $D_{xx} = D_{yy} = 2\,\mathrm{m^2s^{-1}}$, and $D_{xy} = D_{yx} = 0\,\mathrm{m^2s^{-1}}$.

The contaminant is considered to be conservative, in the sense that it is not subject to chemical reactions that lead to its production or consumption. In addition, in the

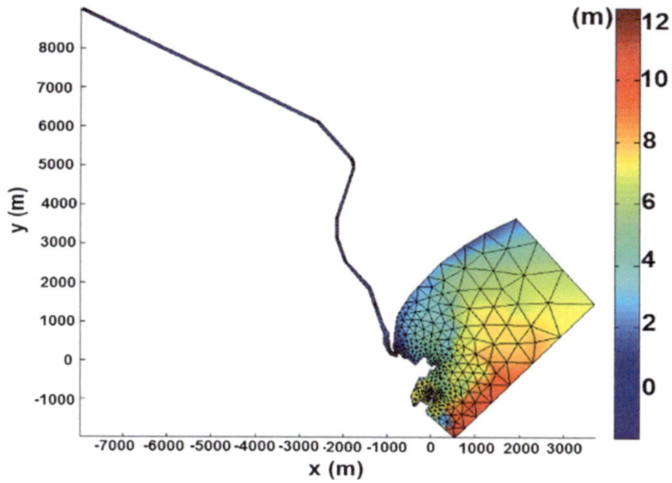

Fig. 5.6 Geometry and bathymetry data (m) of Macaé estuary

fluvial and marine frontiers, the second type, Neumann boundary condition, with zero flux, was used.

$$\frac{\partial C(x, y, t)}{\partial \overrightarrow{n}} = 0 \text{ in } \partial\Omega, \text{for } t > 0 \tag{5.23}$$

As the initial condition, the null concentration of the contaminant in the entire domain of interest is considered, which is compatible with an initially unpolluted region.

$$C(x, y, 0) = 0 \text{ in } \Omega, \text{for } t = 0 \tag{5.24}$$

For the test cases of interest, pollutant punctual and constant in time launches were considered, at specific nodes of the discretized domain.

5.2.5 Hydrodynamical Model

Estuaries subjected to a sufficiently large tidal amplitude, along with a sufficiently small inflow from their main fluvial contribution, may have their hydrodynamics modeled by the two-dimensional formulation known as the Shallow Water Equations (SWE) [17, 114, 127], which solve the flow variables assuming that these vary little in the vertical direction (vertically homogeneous estuary), expressed by

$$\frac{\partial \eta}{\partial t} + \nabla(h\overrightarrow{v}) = 0 \tag{5.25}$$

$$\frac{\partial(\rho hu)}{\partial t} + \nabla\left(\rho uh\vec{v}\right) - \rho fhv + \rho gh\frac{\partial \eta}{\partial x} + \frac{gE_s h^2}{2}\frac{\partial C_{salt}}{\partial x} + \tau_{B_x} - \tau_{W_x} = 0 \quad (5.26a)$$

$$\frac{\partial(\rho hv)}{\partial t} + \nabla\left(\rho vh\vec{v}\right) + \rho fhu + \rho gh\frac{\partial \eta}{\partial y} + \frac{gE_s h^2}{2}\frac{\partial C_{salt}}{\partial y} + \tau_{B_y} - \tau_{W_y} = 0 \quad (5.26b)$$

with [90]

$$E_s = \frac{(\rho_s - \rho_d)}{C_{salt\,max}} \tag{5.27}$$

where η represents the water free surface level above a reference level, corresponding to the mean sea level, $h = \eta + H_0$, see Fig. 4.1, represents the total depth of water, $-H_0$ is the depth of the bottom or bathymetry, ρ is the specific mass of the fluid, f is the coefficient of Coriolis, g is the gravity acceleration constant, τ_{B_x} and τ_{B_y} represent the shear stresses due to the friction with the bottom in the directions x and$y$, respectively, τ_{W_x} and τ_{W_y} represent the shear stresses due to wind force on the water surface in the directions x and$y$, respectively; C_{salt} is the total salt concentration, and E_s, is a dimensionless constant coefficient (mass of the fluid / mass of salt), where $\rho_s = 1{,}025\,\text{kg.m}^{-3}$, $\rho_d = 1{,}000\,\text{kg.m}^{-3}$ and $C_{salt\,max} = 35\,\text{kg.m}^{-3}$.

The hydrodynamic fields used in this chapter were obtained with the MOHID platform [115], considering the following boundary conditions: at the river frontier—river volumetric flow rate of $7.8\text{m}^3\text{s}^{-1}$ (typical condition for the dry season of the Macaé river); and in the maritime border—astronomical tide with 17 components [23], with the amplitude and phases given in Ref. [92]. The simulation period corresponded to 24 h (with a temporal discretization of 20 s), including a tidal wave of syzygy representing the largest sea level in a given period of 28 days.

The correspondence between the model data provided by MOHID and those used in the model described in this chapter, are linked to the spatial and temporal meshes considered, being the former represented in Fig. 5.6. Thus, the hydrodynamic data used are approximated by the data that are closer to the geometry of the reference model.

With the possession of the hydrodynamic variables and the geometry already defined, it is possible to define the time step to be used according to the discretization method. For the combined application of the finite element and finite difference method in the solution of the transport problem, in Ref. [197] it is recommended to make the Courant number as close to unity as possible, for the smallest element of the domain, that is,

$$\nu_{Co} = \frac{\left|\vec{v}\right|\Delta t}{h_e} \tag{5.28}$$

such that

$$\Delta t = \frac{h_e}{|\vec{v}|} \tag{5.29}$$

where h_e corresponds to the length of the smallest side among all elements of the spatial mesh.

Analyzing the smallest element of the estuary geometry, we calculated the smallest Δt according to Eq. (5.29) and, from this, the approximate value Δt=*1,800* s was chosen as the temporal increment, which is a multiple of the time increment of the hydrodynamic model solved in Ref. [92].

5.3 Estimation of Contaminant Source Parameters

A source of contaminants may be characterized either as punctual or distributed, instantaneous, continuous or intermittent, and may also be internal (result of chemical or physicochemical reactions) or external to the domain of interest (such as pollutant discharges). At this point we aim at estimating the location and intensity of a single punctual and constant source. Thus, it is important to understand that the location of the source is a point position (node) in the discrete two-dimensional domain representation of the estuary. Therefore, a different approach, from all other parameters in the analysis is necessary, since the source intensity parameter ($kg\ m^{-3}s^{-1}$) can be sensitive to the size of the specific finite element within which is the estimated location for the source. Besides, the node of this location can also belong to more than one element.

Considering these aspects, the location of the contaminant source must belong to a node of the discretized geometry, and its mass flow rate (kg s^{-1}) must be calculated considering the finite elements to which this node belongs to. As the location of the source is defined by points in the spatial network, and, therefore, referenced to discrete and limited data, to estimate its real location is necessary to use some technique that takes into account only points of the mesh, but at the same time can evaluate such position in relation to the actual estimated location of the source. Given this, there are two interesting perspectives to consider. In one of them, the spatial mesh point is chosen according to the smallest Euclidean distance between the points of the mesh and the point generated by the applied method (projection). In the other, the selection of mesh points belonging to the discretized space with the applied optimization method is performed.

If the estimation of the location of the source considers the projection, it is necessary to evaluate the point estimated by the optimization method used with respect to all points of the mesh. Although there is a certain computational cost, the positive factor is the geometric relation of the estimated point and the chosen point with least distance. This aspect allows some freedom from the method of acting within the initial limits. In contrast, the initial geometric limits imposed together with the calculation of the shortest distance can impose a prevalence of some points of the mesh (see Fig. 5.7), not guaranteeing the same probability of choice for all selected points,

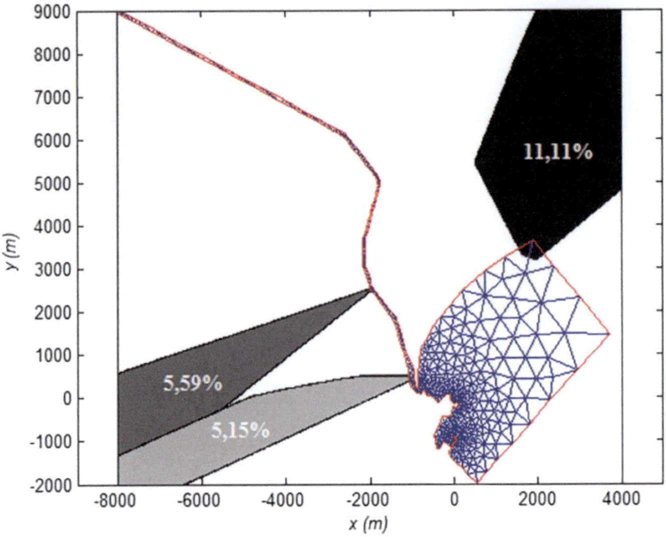

Fig. 5.7 Prevalence of three mesh points in the Nodal Projection Space (NPS)

the objective of most stochastic methods. The points that prevail are those belonging to the borders, which are closer to any area outside the geometry in question. In Fig. 5.7 it is observed the prevalence of three points considering the estimation of points with the projection approach, here called Nodal Projection Space (NPS). The NPS of the three most prevalent points represent 11.11%, 5.59% and 5.15% of the total area considered.

If the estimation of the location of the source considers the selection of the points defined within constraints (geometric relation of neighborhood), it is necessary to evaluate all the points of the mesh to identify the ones that satisfy the constraints. Although there is a certain computational cost, the positive factor is that all selected points have the same probability of being chosen, and, therefore, there is no prevalence of points or regions. On the other hand, it is necessary to define limits (search space) for the selection of the points, which is dependent on the method applied, often requiring adaptation, or even making unfeasible the use of such technique.

5.3.1 *Sensitivity Analysis*

In order to evaluate the possibility of the proposed source parameters identification based on the chosen transport model, it is necessary to analyze their influence using, the so called sensitivity coefficients [42, 99, 124]. If such coefficients are small or indicate the parameters correlation, the estimation problem is difficult and very sensitive to measurement errors [15]. Based on the definition of the modified, or scaled, sensitivity coefficient given in Eq. (1.26), we are interested in calculating

$$X_{L_s}^{i_p,i_t} = L_s \frac{\partial C_{i_p,i_t}}{\partial L_s} \tag{5.30}$$

and

$$X_{F_s}^{i_p,i_t} = F_s \frac{\partial C_{i_p,i_t}}{\partial F_s} \tag{5.31}$$

where L_s is the source location, F_s is the intensity of the source, and i_p ($i_p = 1, 2, \ldots, n_p$), i_t ($i_t = 1, 2, \ldots, n_t$), represents the position and the time of the data acquisition, respectively. We must observe that the parameter F_s is related to the source term S_c in Eq. (5.1). In this chapter we consider that there is no sink term, i.e. $R_c = 0$ in the transport equation.

For the location sensitivity coefficient computation an adaptation is required, since L_s represents the spatial location (node) of the source in the two-dimensional space of the estuary geometry. Thus, it was assumed $L_s = 100$m in Eq. (5.30), and for the numerical derivative approximation,

$$\frac{\partial C}{\partial L_s} \approx \frac{\Delta C}{\Delta L_s} \tag{5.32}$$

where ΔL_s is approximated by the distance of two nodes of the finite element spatial mesh.

Three possible source positions were analyzed separately, considering the mass flow rate of 0.1 kg s^{-1}, with five possible positions of concentration data acquisition collection sensors, i.e. from S_1 to S_5. The locations of the sources and sensors are shown in Fig. 5.8.

The sensitivity coefficients for the three sources proposed, in a total time of 24 h (86,400 s), are shown in Fig. 5.9. For the source located at L_1 (Fig. 5.9a), only sensor S_1 has significant sensitivity for both parameters. As expected, for the source located at L_2 (Fig. 5.9b), the sensor S_1 has no sensitivity whatsoever, not capturing the concentration variation, and the same occurs with the sensor S_5. The sensor S_4 presents a slight sensitivity at the end of the time considered, while the sensors S_2 and S_3 are sensitive for the two analyzed parameters. For the source located at L_3 (Fig. 5.9c), sensor S_1 is not sensitive, sensors S_2 and S_5 present small sensitivities, while sensors S_3 and S_4 have significant sensitivity to both parameters.

Although some sensors do not show sensitivity to some source locations, they may become sensitive to other source locations. This aspect becomes crucial to qualify it as a good location for data acquisition, since in assessing two possible source locations, this distinction may become evident.

In order to evaluate the existence of linear correlation between the analyzed parameters, the Pearson correlation coefficient [108] was calculated for the sensitivity coefficients at each sensor, for each source

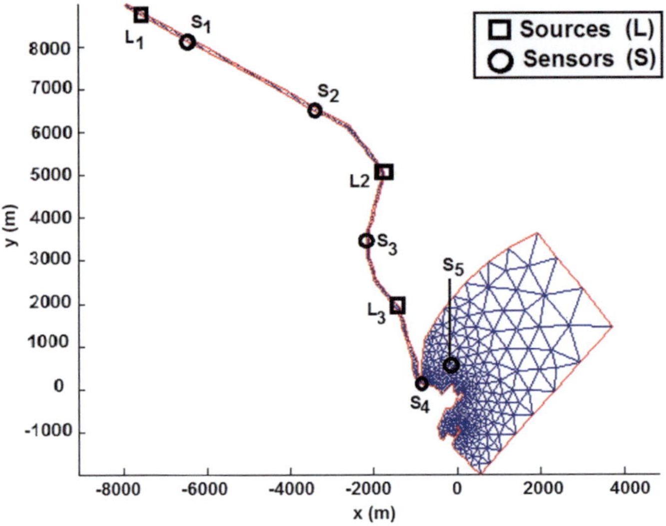

Fig. 5.8 Sources and sensors locations

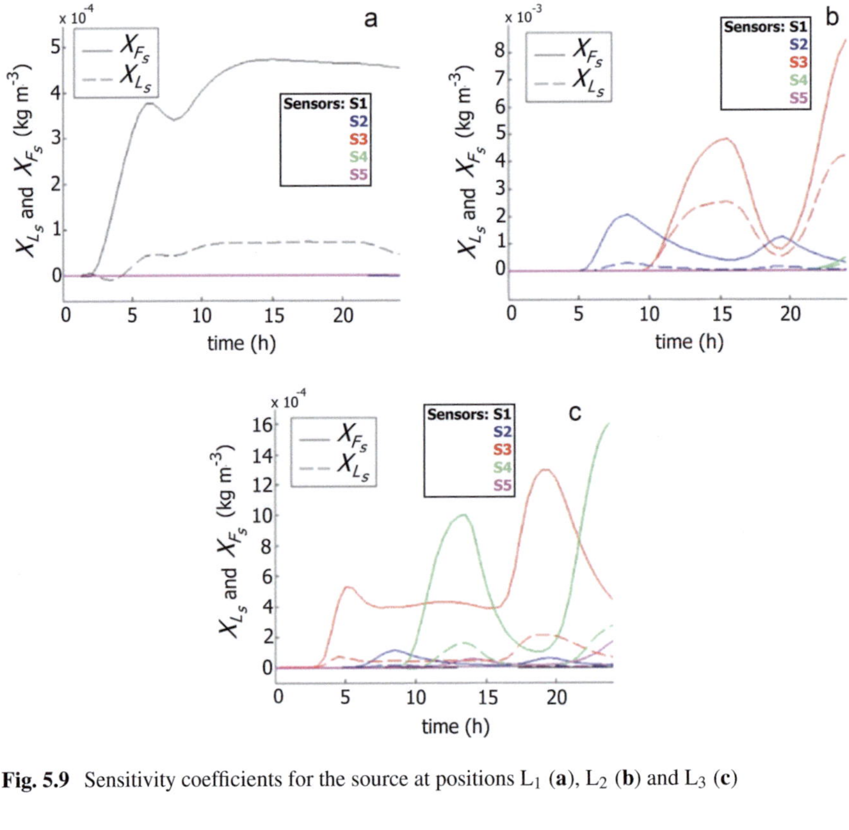

Fig. 5.9 Sensitivity coefficients for the source at positions L_1 (**a**), L_2 (**b**) and L_3 (**c**)

Table 5.2 Pearson coefficients for sources at three locations and sensors at five positions

Source/sensor	S_1	S_2	S_3	S_4	S_5
L_1	0.9335	0.8531	–	–	–
L_2	–	0.9858	0.9940	0.9997	–
L_3	–	0.9998	0.9766	0.9723	0.9721

$$R_P^{i_p} = \frac{\sum_{i_t=1}^{n_t}\left(X_{L_s}^{i_p,i_s} - \overline{X}_{L_s}^{i_t}\right)\left(X_{F_s}^{i_p,i_t} - \overline{X}_{F_s}^{i_p}\right)}{\sqrt{\sum_{i_t=1}^{n_t}\left(X_{L_s}^{i_p,i_t} - \overline{X}_{L_s}^{i_p}\right)^2}\sqrt{\sum_{i_t=1}^{n_t}\left(X_{F_s}^{i_p,i_t} - \overline{X}_{F_s}^{i_p}\right)^2} \tag{5.33}$$

where $\overline{X}_{L_s}^{i_p}$ and $\overline{X}_{F_s}^{i_p}$ represent the arithmetic mean values of the sensitivity coefficients. As shown in Table 5.2, most of the Pearson coefficients are very close to the unity, which shows a strong linear correlation between the location parameters and source intensity. Thus, the process of simultaneous estimation of the source location and intensity may impair the quality of the estimates, since both parameters are correlated. However, it is possible to take two perspectives in the estimation of these parameters: (i) to estimate the location first and then the intensity of the source; and (ii) for each estimate of a source location, to estimate its intensity for this location.

Although the first perspective may seem more feasible, using a constant intensity parameter can generate distinct sources in different locations. Thus, the second perspective becomes more interesting, since at each location, the source intensity parameter adjusts specifically for such location, according to the concentration data. With this, at each point (estimated location) one can employ a one-dimensional estimation algorithm for intensity, which is easy to implement, covering a considerable range of possibilities.

5.3.2 *Experimental Data and Computational Cost*

The experimental data considered here are synthetic data, generated from the solution of the transport problem, i.e. the Direct Problem (DP), using known parameters for the 24 h timeframe of simulation. More details are presented in Sects. 1.32 and 5.42.

The location taken as the real one for the pollutant source was L_2, as shown in Fig. 5.8, with an intensity parameter $F_s = 1.1243 \times 10^{-5}\,\text{kg}\,\text{m}^{-3}\text{s}^{-1}$. The sampling data are the concentration values measured at the sensors S_1, S_2, S_3, S_4, and S_5 locations (see Fig. 5.8) at every hour (i.e. 1, 2, 3, …, 24 h). Thus, the experimental data comprise 120 measurements, acquired at 5 locations and 24 time instants.

All the Direct Problem (DP) and Inverse Problem (IP) programming were developed in C++ language, as well as most of the numerical analyzes performed. The computational cost of each DP execution depends on the location and intensity of the source. For the real scenario, to be estimated, with the known intensity and location, the computational cost of the DP execution is approximately 7.39 s (computer with Intel Core i5 processor, 2.67 GHz and 4 GB of RAM using the Windows 7 Professional operating system).

Considering the source in the position L_1, the approximate computational cost is 4 s, and considering its location in L_3 is 8.61 s. Considering the actual location of the source, L_2, and varying its intensity with the mass flow rates of 0.01 kg s^{-1} and 1 kg s^{-1}, the computational cost corresponds to 6.24 s and 8.64 s, respectively.

According to these variations, although the number of DP executions and computational cost are important, having a direct impact in the IP analysis, one should take them with due caution, since when being smaller, it does not mean obtaining better results.

The Luus-Jaakola method (LJ), the Particle Collision Algorithm (PCA), and the Ant Colony Optimization (ACO) [150] were applied for the estimation of the location of the source. As for the estimation of the location, if the point has already been estimated previously, we move to a new iteration of the applied methods. Estimated a new point, the Golden Section Method (GSM) is then used to estimate the strength of the source at this point. The objective function to be minimized, corresponding to the solution of the IP, is the sum of the squared residues between the calculated and the measured values of concentration. See Eqs. (1.23) and (4.24).

5.4 Results and Discussion

The results obtained for the estimation of the contaminant sources parameters with the application of different optimization methods is presented in this section. For the estimation of the source location, the LJ method, the PCA, and the ACO were applied. For the estimation of the intensity of the source, the GSM was used.

Different stopping criteria may be considered in the IP iterative procedures. For the methods applied in the estimation of the source location, were considered: success in the minimization of the objective function, which should assume a lower value than a prescribed one; and obedience to a maximum number of estimations of different solution candidates (maximum number of evaluations of the objective function). The value of 60 was assumed for such number, which represents 10% of the total number of nodes of the spatial mesh. Regarding the application of GSM, the stopping criterion used was a smaller value than a prescribed one. Each method stores and updates the best result along the iterations/generations.

5.4.1 Estimation of Sources Parameters with Noiseless Experimental Data

In a first application, with noiseless data, the LJ method was used to estimate the location of the source, with projection and selection of the mesh points, and the GSM method to estimate the source intensity.

The choice of the LJ method control parameters was conducted empirically, performing a number of tests in the search of the best configurations. It was then used $n_{out} = 20$ and $n_{int} = 5$ for the number of cycles in the external and internal loops respectively, as well as $\varepsilon = 0.3$ for the contraction coefficient. Therefore, one can estimate up to 100 points. The 30% contraction at each external iteration represents a 51% area reduction in the spatial search domain, which is initially considered as a 10,000 m × 10,000 m square region.

For the GSM, the initial search region [0; 7.7078×10^{-5}] kg $m^{-3}s^{-1}$ was considerd for the source intensity, and 10^{-5} as the tolerance in the stopping criterion. Such interval was chosen to allow the estimation of a value close to the real one for the source intensity parameter that was 1.1243×10^{-5} kg $m^{-3}s^{-1}$ (mass flow rate of 0.1 kgs^{-1}). Under these conditions, seven different intensities are calculated for the chosen location, that is, the direct problem is calculated seven times at each new estimated location.

The solution of the inverse problem with the LJ-GSM approach and the point projection, according to the prescribed conditions, obtained 95% success (20 runs), estimating the source point and the intensity parameter precisely. The solution of the inverse problem using the LJ-GSM and the random estimation of the points within the search space (selection), according to the prescribed conditions, led to the estimation of the source location and intensity parameters precisely in 75% of the total number of executions.

In Figs. 5.10 (projection) and 5.11 (selection), are presented the estimated points for the source location obtained along the LJ method iterative process, as well as the reduction of the search space performed at each external iteration.

As a second application, the PCA method was used to estimate the location of the source, with projection and selection of the mesh points, and the GSM for the source intensity estimation. The choice of the operator parameters of the PCA method was conducted through a number of tests, considering then the best results obtained, resulting in 50 iterations for the external loop, and 10 iterations for the exploitation operator. Besides, the random scattering operator was turned off. The GSM was applied under the same conditions as previously described.

The solution of the inverse problem with the PCA-GSM approach and the points projection strategy resulted in 95% success, considering 20 executions of the method. On the other hand, the strategy with the random estimation of the existing points within the search space (selection), success was achieved only for 50% of the executions. In Figs. 5.12 (projection) and 5.13 (selection) the results obtained with both strategies are shown.

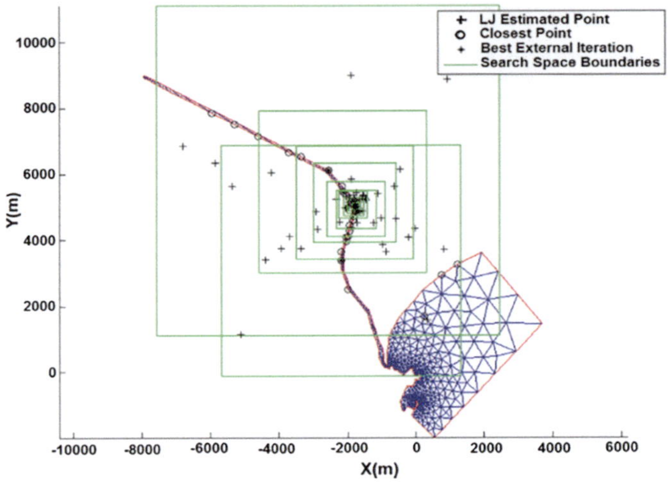

Fig. 5.10 Estimated source location points (projection), and search spaces obtained with the LJ-GSM approach, and noiseless concentration data

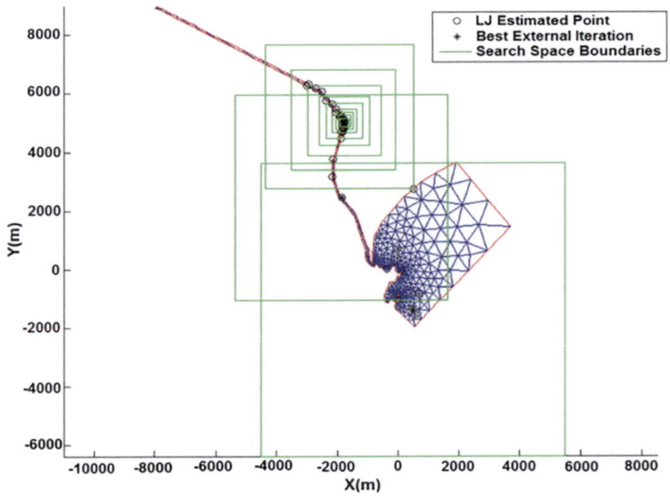

Fig. 5.11 Estimated source location points (selection), and search spaces obtained with the LJ-GSM approach, and noiseless concentration data

Now in a third application, the solution of the inverse problem with the ACO-GSM approach obtained 100% success in the 20 executions that were carried out. In Fig. 5.14 the source location estimated along the iterative process of the ACO method is shown.

As the ACO method deals directly with the nodes in the discretized spatial domain, it corresponds directly to the selection strategy.

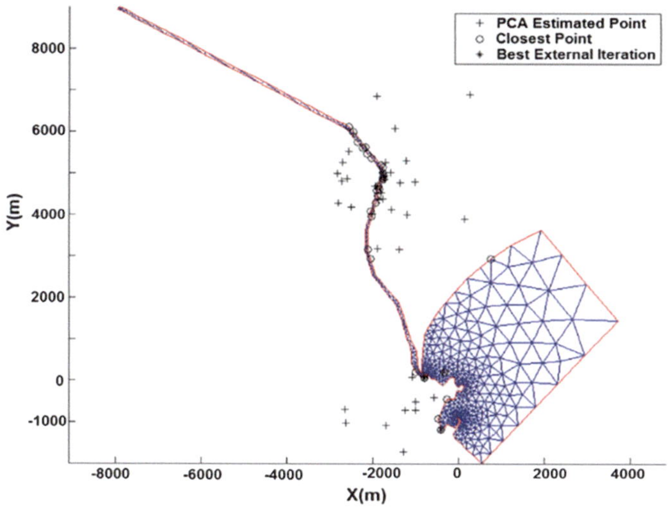

Fig. 5.12 Estimated source location points (projection), obtained with the PCA-GSM approach, and noiseless concentration data

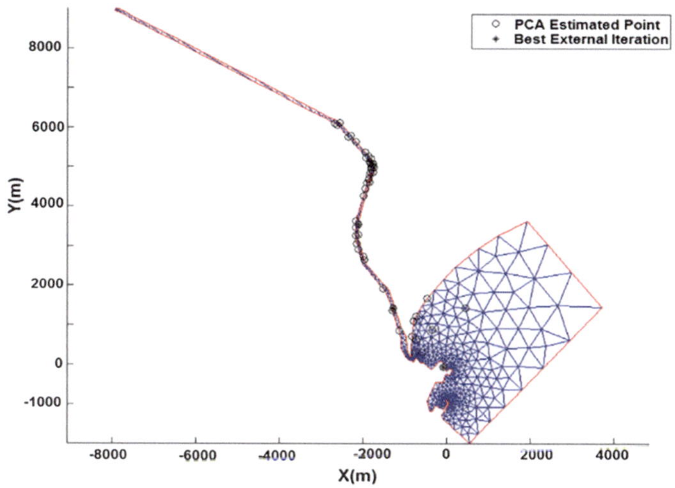

Fig. 5.13 Estimated source location points (selection) obtained with the PCA-GSM approach, and noiseless concentration data

In Table 5.3 the average values from 20 executions are shown for each of the three methods, LJ-GSM, PCA-GSM and ACO-GSM, considering also the projection and selection strategies for the first two methods. In comparison, the use of the point projection strategy presented better results for both LJ-GSM and PCA-GSM, in relation to the selection of the points belonging to the search spaces strategy, showing a difference of 45% for the PCA-GSM. The LJ-GSM and PCA-GSM, with

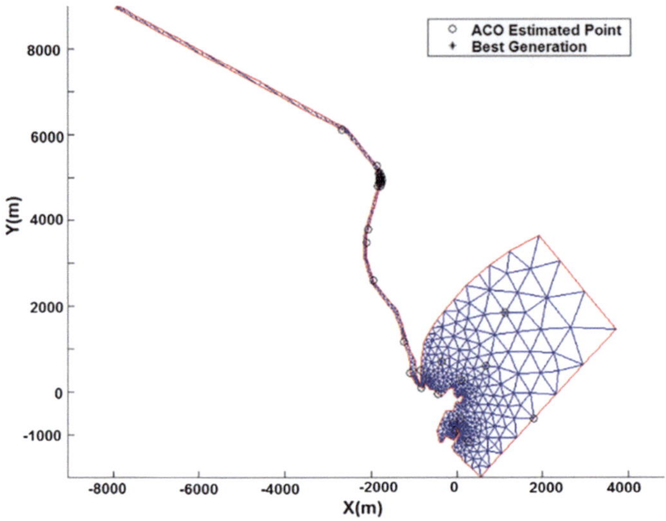

Fig. 5.14 Estimated source location points obtained with the ACO-GSM and noiseless concentration data

the projection strategy and the ACO-GSM were all efficient, with more than 95% accurate executions for both the source location and intensity estimation. The average number of different points estimated for the location (N_{EP}) is similar, i.e. 32.4, 29.8 and 33.6, while the computational cost (time) is slightly higher for the ACO-GSM.

Considering the value of the objective function for the best achieved point along the evolution of the methods, S_b, it is perceived a faster decrease with the LJ-GSM, and that ACO-GSM is the method that requires more iterations to estimate the source

Table 5.3 Performance comparison considering average results in 20 executions of the inverse problem with noiseless concentration data

Methods	Success (%)	Source distance (m)	Source difference (kg s^{-1})*	N_{EP}**	Computational time (s)
LJ-GSM (projection)	95	3.14	6.75×10^{-4}	32.4	1,658.0
LJ-GSM (selection)	75	160.59	-0.0116	38.2	2,221.4
PCA-GSM (projection)	95	3.14	6.75×10^{-4}	29.8	1,585.2
PCA-GSM (selection)	50	439.06	0.0093	42.9	2,513.6
ACO-GSM	100	0	0	33.6	1,926.0

(*) Considering the finite element to which the point belongs, and the corresponding bathymetry
(**) Average number of estimated points

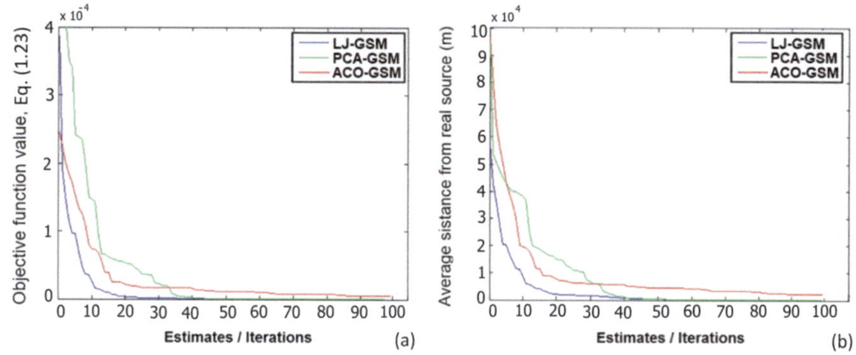

Fig. 5.15 Evolution of the objective function value, Eq. (1.23), for the best achieved point, S_b (**a**), and of the distance from the real source location (**b**), using noiseless concentration data

precisely, as can be seen in Fig. 5.15a. Similar behavior occurs with the average of the actual source distance, shown in Fig. 5.15b.

In order to better compare the methods performance from their results, a nonparametric test, the Wilcoxon signed rank test [41, 61, 75] is used, considering a confidence level of 95%. In Table 5.4 we present the comparison of the performance of the methods regarding the objective function, S_b, the average number of estimated points, N_{EP}, and the computational cost (time). The statistical test results show that the methods do not present significant difference in relation to the adopted criteria, that is, they provided equivalent results for the solution of the proposed inverse problem.

Figure 5.16 shows an analysis of the estimates close to the actual source location, for the intensity and objective function errors, within a distance of less than 200 m from the real source. From each point, the mass fluxes were calculated according to the finite elements to which the point belongs and the corresponding bathymetry, after the application of the GSM for the intensity estimation.

As the distance from the real source increases, the errors do not maintain the growth trend, as might be expected. This fact can be ascertained by the two nearest points of the real source, where the nearest one has a lower Source Intensity Relative Error (SIRE), and higher Objective Function Relative Error (OFRE). As the inverse problem solution aims at minimizing the objective function, the second point seems to be more apt to solve the problem. However, the best estimate is the third closest point (62.87 m), since it has the lowest OFRE value, and also the lowest SIRE (13.5%).

Table 5.4 Wilcoxon signed rank test results for the inverse problems solutions with noiseless concentration data

Comparison criterion	Methods compared	R +	R−	Critical value	*p* Value	Significant difference
S_b	LJ-GSM *x* PCA-GSM	110	110.5	52	1	No
	LJ-GSM *x* ACO-GSM	115	95	52	1	No
	PCA-GSM *x* ACO-GSM	115	95	52	1	No
N_{EP}	LJ-GSM *x* PCA-GSM	125.5	84.5	52	0.4436	No
	LJ-GSM *x* ACO-GSM	91	121	52	0.6007	No
	PCA-GSM *x* ACO-GSM	79.5	130.5	52	0.3543	No
Time	LJ-GSM *x* PCA-GSM	113	97	52	0.7652	No
	LJ-GSM *x* ACO-GSM	60	150	52	0.0930	No
	PCA-GSM *x* ACO-GSM	61	149	52	0.1005	No

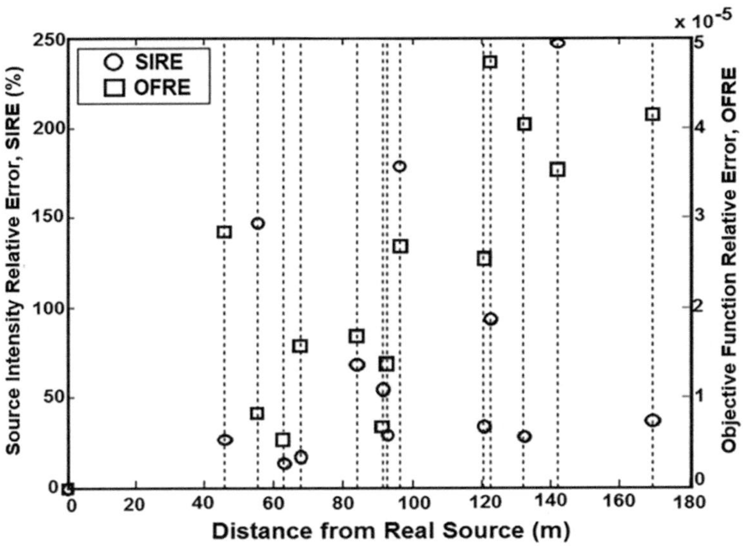

Fig. 5.16 Source Intensity and Objective Function Relative Errors, SIRE and OFRE, respectively, for the estimates closer to the actual pollutant source location, for the inverse problem solutions with noiseless concentration data

5.4.2 Estimation of Sources Parameters with Noisy Experimental Data

In real situations, with experimentally collected data, there are always errors which are inherent to the equipment used, experimental design setup, conditions and simplifications, besides human errors in the experiment realization. The data used in the previous section was considered noiseless, consisting on the calculated values for the concentration using the direct problem model. Here, using also synthetic data, we try to reproduce a more realistic situation in order to test the robustness of the estimation methods, by adding a pseudorandom error of the order of $\pm$ 5%, calculated by (see Eq. (1.27))

$$C_{exp_i} = C_{calc_i} + \sigma_{exp} r_i, \; for \; i = 1, 2, ..., N_d \tag{5.34}$$

where C_{exp_i} represents the synthetic noisy concentration experimental data, C_{calc_i} represents the calculated values for the concentration, using the exact values for the unknowns we intend to estimate with the solution of the inverse problem, $N_d = n_p \times n_t$ is the total number of experimental data, r_i is a random number in the range $[-1,1]$, and the experimental measurement error standard deviation was calculated with$\sigma_{exp} = 0.05 \times max(C_{calci})$, representing, therefore, a maximum experimental noise, of the order of 5%.

For very small or negative concentration values, the noise application was disregarded, that is, if $C_{calc_i} < 10^{-5}$, then $r_i = 0$ or if $C_{exp_i} < 0$, it was replaced by $C_{exp_i} = C_{calc_i}$.

The noiseless and the 5% noise level experimental data are presented in Fig. 5.17, for the five sensors considered, as a function of time.

In the methods used for the inverse problem solution with noisy data, only the projection strategy was considered, as it led to better results in the previously shown applications. The LJ, PCA, ACO and GSM control parameters used were the same as in the previous applications. The solution of the inverse problem with both the LJ-GSM and PCA-GSM, and 5% level noisy data, achieved 80% success in estimating the source with good accuracy.

Figures 5.18 and 5.19 present the points estimated along the iterative process of the LJ and PCA methods, respectively, using the projection strategy, and data with 5% noise level.

The solution of the inverse problem with the ACO-GSM and data with 5% of noise level, obtained 80% success in the 20 executions carried out. In Fig. 5.20 are shown the points estimated along the iterative process of the ACO-GSM application.

Table 5.5 shows the average values from 20 executions for each of the three methods used in the inverse problem solution, considering 5% noise level concentration data. The PCA-GSM presented a smaller distance from the actual source location, but the source intensity was the one with the largest difference. The LJ-GSM reached the lowest number of estimated points for the source location, N_{EP}, and also the lowest computational time in relation to the other methods.

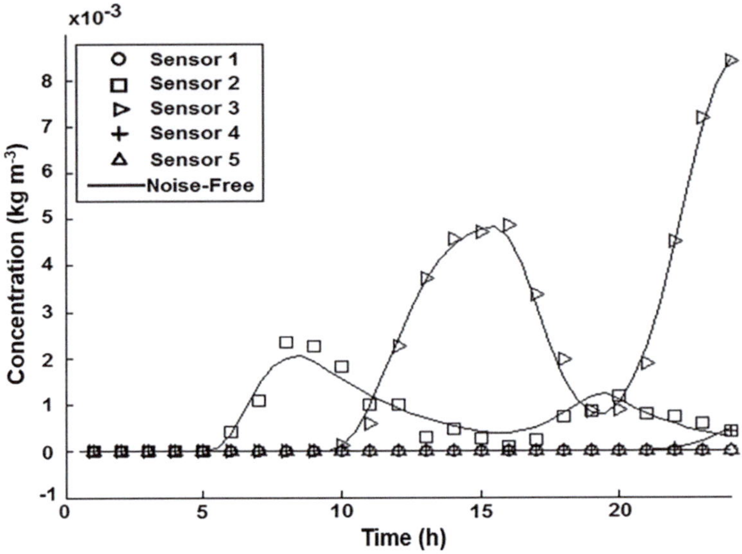

Fig. 5.17 Noiseless and 5% noise level concentration experimental data

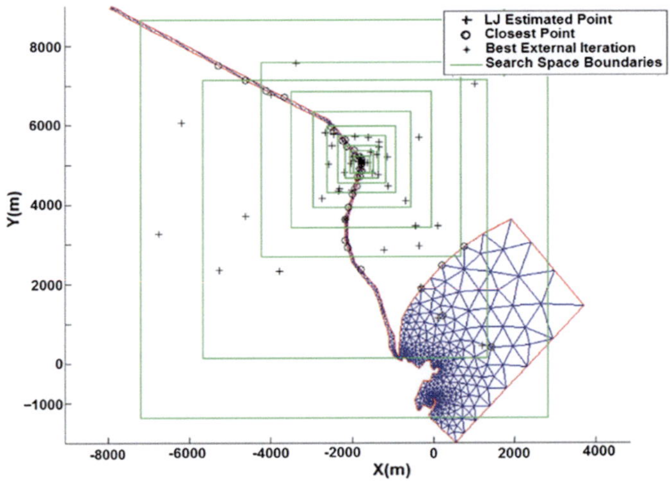

Fig. 5.18 Estimated source location points (projection), and search spaces obtained with LJ-GSM approach and concentration data with 5% noise level

Considering the value of the objective function for the best achieved point along the evolution of the methods, the PCA-GSM shows a similar behavior to the LJ-GSM and ACO-GSM in the first iterations, but after that decreases faster, as shown in Fig. 5.21a. The same behavior is observed for the average distance from the actual source location, which can be observed in Fig. 5.21b.

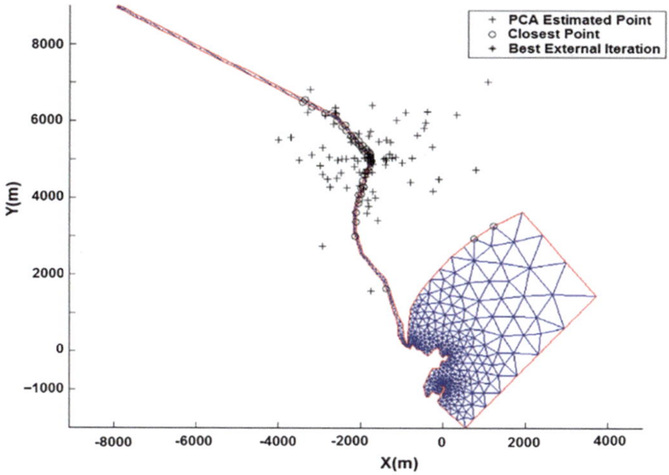

Fig. 5.19 Estimated source location points (projection) obtained with the PCA-GSM approach and concentration data with 5% noise level

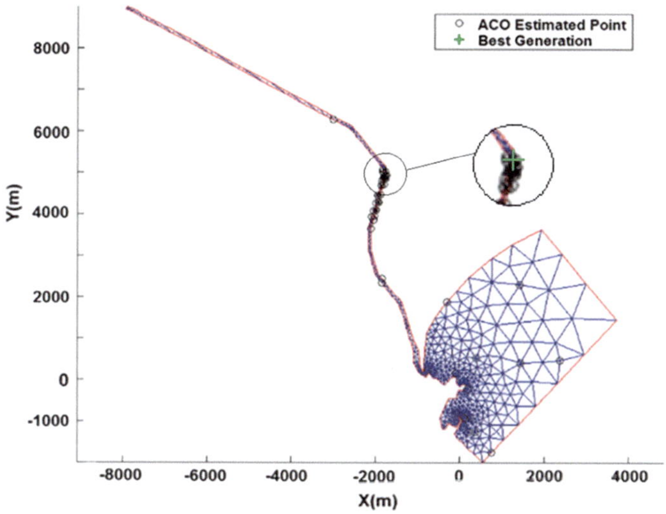

Fig. 5.20 Estimated source location points with the application of the ACO-GSM approach using concentration data with 5% noise level

The Wilcoxon signed rank test with 95% confidence level was used for the comparison of the methods results with respect to the objective function value of the best achieved estimated point, S_b, the number of estimated points, N_{EP}, and the computational cost (time). The results are shown in Table 5.6. It can be observed that even using data with 5% noise level, the results for S_b do not present significant difference

Table 5.5 Performance comparison considering average results in 20 executions of the inverse problem with 5% noise level concentration data

Methods	Success (%)	Source distance (m)	Source difference (kg s^{-1})	N_{EP}	Computational time (s)
LJ-GSM	80	805.34	0.0096	31.65	1,653.6
PCA-GSM	80	12.52	0.0187	44.20	2,322.8
ACO-GSM	80	201.71	0.0065	40.05	2,299.2

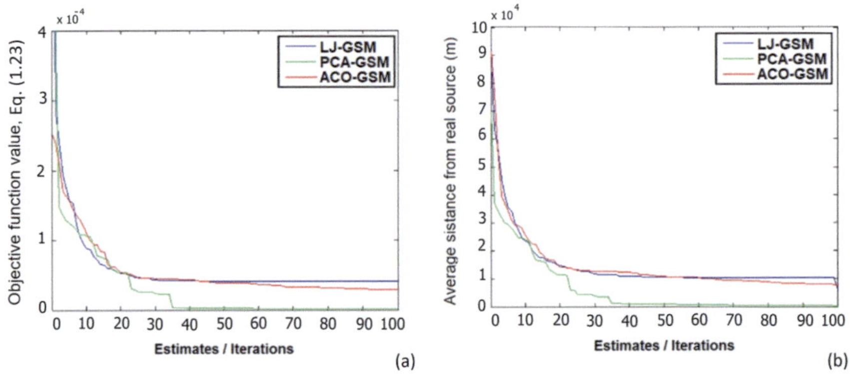

Fig. 5.21 Evolution of the objective function value, Eq. (1.23), for the best achieved point, S_b (**a**), and of the distance from the real source location (**b**), using 5% noise level concentration data

for the three methods studied. Regarding the performance, based on the number of estimated points (N_{EP}) and time, the LJ-GSM was superior to the others.

Figure 5.22 shows an analysis of the estimates close to the actual source location for the intensity and the objective function errors, within a distance of less than 200 m from the real source. As in the case with noiseless data, when using concentration data with 5% noise level, points closer to the actual source location may be less suitable than those located at larger distances. From the analyzed points, the third closest point (62.87 m) is the best estimate for the inverse problem solution, since it presents the lowest source intensity relative error (SIRE) (13.5%), and one of the lowest objective function relative error (OFRE).

5.5 Remarks

In the estimation of the location and intensity of a hypothetical punctual and continuous pollutant discharge, in the estuarine region of the Macaé River, it was verified, through a sensitivity analysis, that these parameters are correlated, which makes

Table 5.6 Wilcoxon signed rank test results for the inverse problems solutions with 5% noise level concentration data

Comparison criterion	Methods Compared	R+	R−	Critical value	*p* value	Significant difference
S_b	LJ-GSM ***x*** PCA-GSM	112.5	99.5	52	0.5234	No
	LJ-GSM ***x*** ACO-GSM	116	94	52	0.5469	No
	PCA-GSM ***x*** ACO-GSM	104	108	52	0.7188	No
N_{EP}	LJ-GSM ***x*** PCA-GSM	36.5	173.5	52	0.0105	Yes
	LJ-GSM ***x*** ACO-GSM	35.5	174.5	52	0.0089	Yes
	PCA-GSM ***x*** ACO-GSM	135.5	74.5	52	0.2758	No
Time	LJ-GSM ***x*** PCA-GSM	43	167	52	0.0206	Yes
	LJ-GSM ***x*** ACO-GSM	26	184	52	0.0032	Yes
	PCA-GSM ***x*** ACO-GSM	111	99	52	0.8228	No

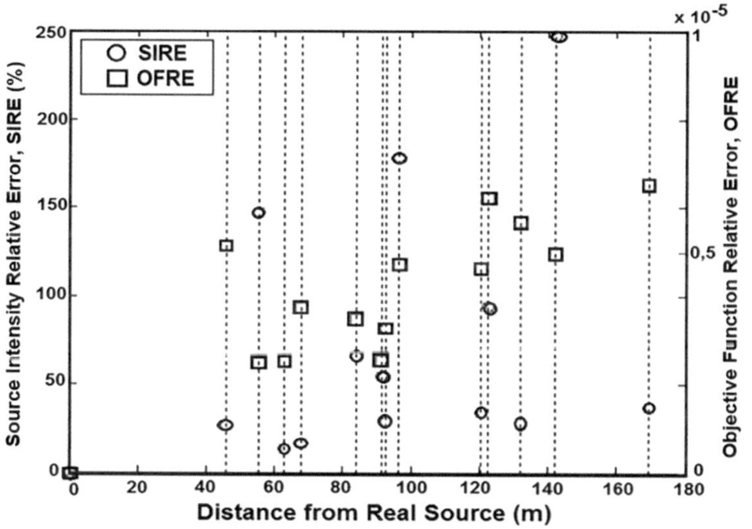

Fig. 5.22 Source Intensity and Objective Function Relative Errors, SIRE and OFRE, respectively, for the estimates closer to the actual pollutant source location, for the inverse problems solutions with 5% noise level concentration data

it difficult to apply inverse problem techniques. The source intensity parameter (kg $m^{-3}s^{-1}$) was sensitive to the domain geometry through the finite element dimensions resulting from the discretization process. Therefore, it was alternatively chosen to estimate the location of the source, and during this process, for each specific location, to estimate the intensity parameter. In this way, the intensity parameter is adjusted according to the location of the source, and therefore, to the geometry and water level considered.

Considering three different locations of the source, five sampling positions (sensors) were proposed, and the sensitivity of these sensors to the position of the source was verified. Thus, even if a sensor is not sensitive to the real source, having sensitivity to a given estimated location, it was induced the solution of the problem to errors of calculation of concentrations, resulting in the elimination of the estimate itself as a solution of the inverse problem.

As the location of the source was defined through a node of the discrete geometry of the Macaé River estuary, resulting from the Finite Element Method domain discretization, the estimation of this location had to contemplate these nodes as possible locations. In addition, because the discretized geometry in the domain of interest presented irregularities that prevented the framing in a well-defined search space, two strategies were applied, so that the solution methods of the inverse problem estimated only nodes of the spatial mesh. One of them makes the projection of points, that is, when a point is estimated in a search space, this point is projected to the node closest to the mesh, making it the estimated point. The other strategy made the selection of nodes belonging to the search space, and then, the choice of the points was performed randomly among them. Each strategy has its advantages and disadvantages, with respect to the geometry and the methods of solution of the inverse problem. Using the Luus-Jaakola method and the Particle Collision Algorithm, the strategy that proved most efficient for the estimation of the location of the source was the projection technique. In this way, it was chosen to perform the subsequent applications of the inverse problem.

Using noiseless synthetic concentration data, the applied methods had comparable performances, using the projection technique, when possible, to estimate the location and intensity of the source. All methods accurately estimated the source parameters in at least 95% of runs. In terms of the objective function, the number of points estimated for the source location and the computational cost, the three methods applied, LJ-GSM, PCA-GSM and ACO-GSM, are equally competitive for the solution of the proposed inverse problem.

With the use of 5% noise in the concentration data, the three methods showed a 80% success rate for the accurate estimation of the source parameters. Compared based on the objective function, the methods showed no significant difference. However, when compared using the number of different estimated points and computational cost, the LJ-GSM proved to be superior, with better performance in these criteria.

The adaptations made in the canonical version of the ACO method to solve the problem of estimation of the sources, especially with regard to the location, proved to be potentially efficient when compared to the other methods applied for the solution of the inverse problems with noiseless data. And even with the introduction of noise of the order of 5%, although it did not present the best result, the method presented good results, comparable to that of the PCA method.

Applying the GSM for the intensity estimation and analyzing possible estimates with a location close to the real source (distance < 200 m), it was concluded that the errors do not decrease monotonically with the approximation of the estimated source in relation to the source real location. Thus, it was observed that the third closest point yields the best estimate for the solution of the inverse problem, with both the use of noiseless and 5% noise level concentration data. That can be explained by the fact that this estimate was able to reconcile a smaller error between the experimental and calculated concentration values with a smaller error in the estimation of the intensity of the source.

Finally, the use of inverse problems formulation and solution for the estimation of the location and intensity of a source that reproduces the punctual and continuous discharge of a contaminant in estuarine environment, has a strong potential of application in the management of these bodies of water, in decision-making or even in the identification of possible environmental liabilities.

Chapter 6
Estimation of Parameters in a Non-cohesive Sediment Transport Model in Rivers and Estuaries

In this chapter an analytical–numerical hybrid solution is developed through the Generalized Integral Transform Technique (GITT) for a two-dimensional transient problem of transport of non-cohesive sediments. To accelerate the convergence of the hybrid solution, analytical filtering was applied in order to homogenize the boundary conditions, and the equation was transformed into a coupled system of one-dimensional partial differential equations solved numerically employing the computational platform Wolfram Mathematica 13.2 [193]. The results were compared with the solution obtained by Mathematica's routine NDSolve, indicating the feasibility of the solution of this problem using the GITT, which led to a lower computational cost. In order to solve the inverse problem, Bayesian inference was used to identify the dispersion coefficients and the friction and sedimentation velocities, using the Markov Chain Monte Carlo method (MCMC) with the Metropolis–Hastings algorithm to generate the posteriori probability distributions. With a sensitivity analysis it was possible to interpret all the results obtained in a simple way, and to develop efficient experimental tools to obtain accurate estimates for the parameters of interest. Good results have been achieved, and MCMC demonstrated significant efficacy in the identification of parameters associated to the transport and dispersion of non-cohesive sediments in natural flows.

Ademilton Luiz Rodrigues de Souza, from CEFET/RJ, Federal Center of Technological Education, and Diego Campos Knupp, from IPRJ-UERJ, Polytechnic Institute, Rio de Janeiro State University, co-authored this chapter.

J. Lugon Junior et al., *Inverse Problems in the Environmental Modelling of Rivers and Estuaries*, Studies in Systems, Decision and Control 246,
https://doi.org/10.1007/978-3-032-09937-2_6

6.1 Introduction

Over the years there has been a growing concern about the increased concentration of pollutants in bodies of water, which are mainly originated by disordered industrial and population growth. The movement of sediments and pollutants represent one of the most common problems faced by the world population, and for this reason the knowledge and characterization of the sediment behavior in the water environment are essential for the protection of the environment, maintenance of navigation channels, hydraulic works construction, soil conservation, and water resources availability [160]. Global population growth, increasement of water, energy and food demand, and climate change are the most important challenges today to our society [21, 89].

For a long time, the only available solution methods for the advection–dispersion problems were purely analytical techniques. Only in the recent decades emerged the development of large-scale application numerical methods due to the lower cost and wider availability of high-performance computers [88]. Numerical models are an important alternative for the qualitative and quantitative analysis of the hydro sedimentological processes that occur in water bodies, which with the advancement of computing has been gaining increasing interest in engineering projects [137].

In Refs. [38, 39, 51, 63] analytical–numerical hybrid models are used to represent the advection–dispersion processes that govern the transport of pollutants in rivers. In order to find new alternatives for solving differential equations, and to reduce computational cost through systematic approaches, different methodologies have been presented in the literature aiming at combining the insights provided by the analytical methods with the flexibility of numerical methods. The Generalized Integral Transform Technique (GITT) is a hybrid method that has been applied successfully in a wide variety of problems, being able to generate quite efficient computational solutions, not requiring upfront the discretization of the domain of interest, in addition to maintaining automatic control over the error [38].

The use of hybrid methods has become even more interesting and viable with the evolution of programming systems. Along that, the use of the Mathematica platform [193], a state-of-the-art software that brings together symbolic computing resources, enables the user to develop programs with automatic manipulation of analytical expressions. In addition to being recognized, both for its technical capacity and for its easy and sophisticated programming, this platform provides an integrated system, in continuous expansion with several libraries to be used in the most different areas of knowledge. In the literature, several works are presented with proposals for solutions of simplified versions of the advection–dispersion equation that simulates the transport of non-cohesive sediments [161, 163]. The objective of this chapter is to present a hybrid solution for the 2D-vertical model of transport of non-cohesive sediments using the Generalized Integral Transform Technique (GITT), and estimate hydrodynamic and transport parameters present in such model by means of the Bayesian inference.

6.2 Formulation of the Direct Problem

The concentration of non-cohesive sediments, for each granulometric class, can be represented by the following advection–dispersion equation, and respective boundary and initial conditions [129, 132]:

$$\frac{\partial C}{\partial t} + U(z)\frac{\partial C}{\partial x} - W_s\frac{\partial C}{\partial z} = D_x\frac{\partial^2 C}{\partial x^2} + D_z\frac{\partial^2 C}{\partial z^2} \tag{6.1a}$$

$$C(0, z, t) = C_0 \tag{6.1b}$$

$$\frac{\partial C(x, z, t)}{\partial x}\Big|_{x=L_x} = 0 \tag{6.1c}$$

$$W_s C(x, L_z, t) + D_z\frac{\partial C(x, z, t)}{\partial z}\Big|_{z=L_z} = 0 \tag{6.1d}$$

$$W_s C_a + D_z\frac{\partial C(x, z, t)}{\partial z}\Big|_{z=0} = F_D - F_E \tag{6.1e}$$

$$C(x, z, 0) = C_1 \tag{6.1f}$$

where $U(z)$ represents the velocity of the flow in the x direction, with dependence on the vertical coordinate z, W_s is the velocity of the sediment particles fall, D_x and D_z are the turbulent diffusion coefficients in the x and z directions, respectively, L_x and L_z are the longitudinal and vertical lengths of the domain of interest, C_0 represents the continuous emission of a pollutant at the domain boundary, at the position $x = 0$, C_1 is the initial sediment concentration, C_a is a reference concentration in the flow bed under equilibrium conditions, and F_D and F_E are the sediment deposition and erosion fluxes, respectively.

The longitudinal velocity $U(z)$, can be determined by considering the logarithmic law of the wall [90],

$$\frac{U(z)}{u^*} = \frac{1}{\kappa^*} ln\left(\frac{z}{z_0}\right), z > 0 \tag{6.2}$$

where $\kappa^* = 0.4$ is the von Karman constant, z_0 is a constant corresponding to the layer from the bed where velocity is zero, i.e. the distance from the bottom at which the velocity given by the law of the wall goes to zero, and u^* is the friction velocity.

6.3 Generalized Integral Transform Technique

Cotta [38] developed the formalism of the Generalized Integral Transform Technique (GITT) in which the solution of a problem can be obtained by applying an integral transformation pair, generating an auxiliary system incorporating analytical characteristics of the original problem, and a simple infinite coupled system of numerically solved partial or ordinary differential equations, which is truncated at a sufficiently large order, N_{tr}, , to meet the precision requirements of the solution, and an inversion formula is then used to obtain the analytical–numerical hybrid solution [13, 24, 26, 120, 161].

The solution of the problem of interest in this chapter is obtained with the GITT technique, by a partial transformation scheme, only in the vertical direction, resulting in a transformed problem formed by a system of one-dimensional transient partial differential equations. For the problem presented, an analytical filter solution was proposed to remove the information from the source terms present in the boundary conditions, in the z-direction, in order to accelerate the convergence of the expansions in eigenfunctions resulting from the eigenvalue problem. In this way, it was considered that the concentration field $C(x, z, t)$ of the desired solution can be represented by the sum of a filtered potential and a filter function, as shown below:

$$C(x, z, t) = C^*(x, z, t) + C_f(z) \tag{6.3}$$

where $C^*(x, z, t)$ is the filtered potential obtained by the GITT technique, as an expansion in eigenfunctions, and $C_f(z)$ is the solution for the proposed filter.

The function $C_f(z)$ is found by means of the following second-order filtering problem, obtained by differentiating Eq. (6.3) and substituting it in Eqs. (6.1a–6.1f):

$$-W_s\frac{\partial C_f(z)}{\partial z} = D_z\frac{\partial^2 C_f(z)}{\partial z^2} \tag{6.4a}$$

$$W_s C_f(L_z) + D_z\frac{\partial C_f(z)}{\partial z}\Big|_{z=L_z} = 0 \tag{6.4b}$$

$$W_s C_a + D_z\frac{\partial C_f(z)}{\partial z}\Big|_{z=0} = F_D - F_E \tag{6.4c}$$

The filter problem given by Eqs. (6.4a–6.4c) is solved analytically, with the solution given by the following expression:

$$C_f(z) = \frac{exp\left[-\frac{W_s}{D_z}z\right](-F_D + F_E + W_s C_a)}{W_s} \tag{6.5}$$

Once the filter expression is known, the filtered problem can be formulated, so that it is homogeneous in the boundary as follows:

$$\frac{\partial C^*(x,z,t)}{\partial t} + U(z)\frac{\partial C^*(x,z,t)}{\partial x} - W_s\frac{\partial C^*(x,z,t)}{\partial z} = D_x\frac{\partial^2 C^*(x,z,t)}{\partial x^2} + D_z\frac{\partial^2 C^*(x,z,t)}{\partial z^2} \tag{6.6a}$$

$$C^*(0,z,t) = C_0 - C_f(z) = 0 \tag{6.6b}$$

$$\frac{\partial C^*(x,z,t)}{\partial x}\bigg|_{x=L_x} = 0 \tag{6.6c}$$

$$W_s C^*(x,L_z,t) + D_z\frac{\partial C^*(x,z,t)}{\partial z}\bigg|_{z=L_z} = 0 \tag{6.6d}$$

$$\frac{\partial C^*(x,z,t)}{\partial z}\bigg|_{z=0} = 0 \tag{6.6e}$$

$$C^*(x,z,0) = C_1 - C_f(z) \tag{6.6f}$$

The GITT approach can be applied using a partial transformation methodology, when only one spatial coordinate is eliminated by integral transformation. Such strategy uses more robust numerical methods, and therefore can reduce the computational cost by algebraic manipulation and obtain more accurate solutions [27, 39].

For the solution of the problem addressed in this chapter we considered a partial transformation scheme, only in the vertical direction, to eliminate the variable z. . By the simplicity of the boundary conditions present in the problem [39], the spatial coordinate x is not eliminated by a second integral transformation. This procedure results in a transformed problem, thus formed by a coupled system of one-dimensional partial differential equations for the transformed potentials, which can be numerically treated to complete the solution procedure. This combination of solution methodologies is justified by providing an analytical and low-cost computational representation for the expansion in eigenfunctions for the vertical coordinate together with a flexible and feasible numerical approach for the longitudinal behavior of the coupled transform potential [27, 28].

Following the procedure for integral transformation, the eigenvalue problem is chosen so that it contains the maximum information about the problem addressed. For this, we used the direct application of the method of separation of variables in the problem, given by Eqs. (6.1a–6.1f), and the following eigenvalue problem is proposed:

$$D_z\frac{d^2\psi_i(z)}{dz^2} + \beta_i^2\psi_i(z) = 0 \tag{6.7a}$$

$$W_s C_a + D_z\frac{d\psi_i(0)}{dz} = 0 \tag{6.7b}$$

$$W_s \psi_i(L_z) + D_z \frac{d\psi_i(L_z)}{dz} = 0 \tag{6.7c}$$

where β_i are the eigenvalues and $\psi_i(z)$ their respective eigenfunctions.

The problem investigated allows to define the following integral transformation pair

$$\text{Transform : } \tilde{C}_i(x,t) = \int_0^{L_z} \widetilde{\psi}_i(z) C^*(x,z,t) dz \tag{6.8}$$

$$\text{Inverse : } C^*(x,z,t) = \sum_{i=1}^{\infty} \widetilde{\psi}_i(z) \tilde{C}_i(x,t) \tag{6.9}$$

where $\tilde{C}_i(x,t)$ are the transformed potentials, and $\widetilde{\psi}_i(z)$ are the normalized eigenfunctions, whose expression for the latter is given by

$$\widetilde{\psi}_i(z) = \frac{\psi_i(z)}{\sqrt{N_i}} \tag{6.10a}$$

$$N_i = \int_0^{L_z} \widetilde{\psi}_i^2(z) dz \tag{6.10b}$$

Applying the operator $\int_0^{L_z}(.)\widetilde{\psi}_i dz$ in Eqs. (6.6a–6.6f) the following system of partial differential equations (PDE) is obtained

$$\frac{\partial \tilde{C}_i(x,t)}{\partial t} + \sum_{j=1}^{\infty} \frac{\partial \tilde{C}_j(x,t)}{\partial x} \int_0^{L_z} \widetilde{\psi}_i(z) \widetilde{\psi}_j(z) U(z) dz -$$

$$W_s \sum_{j=1}^{\infty} \tilde{C}_j(x,t) \int_0^{L_z} \widetilde{\psi}_i(z) \frac{d\widetilde{\psi}_j(z)}{dz} dz = D_x \frac{\partial^2 \tilde{C}_i(x,t)}{\partial x^2} - \beta_i^2 \tilde{C}_i(x,t) \tag{6.11a}$$

$$\tilde{C}_i(0,t) = \int_0^{L_z} \widetilde{\psi}_i(z) \big(C_0 - C_f(z)\big) dz \tag{6.11b}$$

$$\frac{\partial \tilde{C}_i(L_x,t)}{\partial x} = 0 \tag{6.11c}$$

$$\tilde{C}_i(x,0) = \int_0^{L_z} \widetilde{\psi}_i(z) \big(C_1 - C_f(z)\big) dz \tag{6.11d}$$

The above system of PDEs is solved numerically for the transformed potentials $\tilde{C}_i(x, t)$, so that the inversion formula, given by Eq. (6.9), can be used to obtain the solution for the filtered potential, $C^*(x, z, t)$.

The implemented solution for the direct problem of interest requires the knowledge of parameters that are difficult to measure, such as the longitudinal, D_x, and vertical, D_z, turbulent diffusion coefficients. A very efficient strategy for the estimation of these two dispersive parameters, as well as for the sediment particle settling velocity, W_s, and the friction velocity near the bed bottom, u^*, can be obtained by analyzing the inverse problem.

As shown in Sect. 3.2, Eq. (3.35), the infinite sum in Eq. (6.9) and, therefore, in Eq. (6.11a), is truncated with a finite number of terms N_{tr}, referred to as truncation order.

6.4 Formulation of the Inverse Problem

Evidently, the deterministic models made a great contribution to the understanding of the dynamic behavior of a system. However, do not provide a natural treatment of the uncertainties present in a model, and, therefore, may end up being insufficient in the stages of decision making. Measurement errors are not the only factors that contribute to these uncertainties along the inverse problem estimation process. Other factors, such as the formulation of the problem and the corresponding boundary conditions, as well as the technique used to optimize the unknown parameter vector, may affect the final solution, and present an inconsistent result in the inverse analysis, even if at first all the variables seem to be well defined. Furthermore, there is a diversity of physical phenomena that can be described through stochastic processes due to their ability to simulate noise and uncertainties present in experimental observations [74].

In this chapter, the inverse problem we are interested in consists in obtaining estimates for the parameters D_x, D_z, u^* and W_s, present in Eqs. (6.1a–6.1f) and (6.2). For that purpose, it is used Bayesian Inference. The advantages of this approach are the possibility of including a priori information, and incorporating it into a formal decision context, as well as the explicit treatment of uncertainties, besides the ability to assimilate new information in adaptive contexts [63].

The solution of the inverse problem of parameter estimation with the Bayesian approach adopted here consists in obtaining the posterior probability for the vector of unknowns, $\boldsymbol{P}$, $\pi_{pos}(\boldsymbol{P})$, with the use of Bayes' theorem, as already given by Eq. (3.37), which is reproduced here:

$$\pi_{pos}(\boldsymbol{P}) = \pi(\boldsymbol{P}|\boldsymbol{Y}) = \frac{\pi(\boldsymbol{Y}|\boldsymbol{P})\pi_{pr}(\boldsymbol{P})}{\pi(\boldsymbol{Y})} \tag{6.12}$$

where $\pi_{pr}(\boldsymbol{P})$ is the priori probability, $\pi(\boldsymbol{Y})$ is the marginal probability, and $\pi(\boldsymbol{Y}|\boldsymbol{P})$ is the likelihood, which describes the probability of finding the experimental data $\boldsymbol{Y}$, as already given by Eq. (3.38), which is also reproduced here:

$$\pi(\boldsymbol{Y}|\boldsymbol{P}) = \frac{1}{\sqrt{(2\pi)^{N_d}}} \frac{1}{\det(\boldsymbol{W})} \times exp\left[-\frac{1}{2}(\boldsymbol{Y} - \boldsymbol{Y}_{calc}(\boldsymbol{P}))^T \boldsymbol{W}^{-1} (\boldsymbol{Y} - \boldsymbol{Y}_{calc}(\boldsymbol{P}))\right] \tag{6.13}$$

where $\boldsymbol{W}$ is the covariant matrix of the experimental data error, N_d is the number of experimental data considered, and $\boldsymbol{Y}_{calc}$ is the vector with the values calculated with the direct problem solution.

The a priori information cannot always be represented by a normal, or even a uniform, distribution, which makes it very difficult to obtain the a posteriori probability distribution. In such cases, it is necessary to use sampling techniques to simulate the posterior distribution, and then infer measures of central tendency and dispersion of this distribution [74]. Among these sampling techniques, the Markov Chain Monte Carlo method (MCMC) stands out. The essential idea of MCMC is to simulate a random sequence of the posterior distribution for each parameter of interest of the inverse problem, converging to a stationary (equilibrium) distribution.

One of the most used algorithms for the generation of Markov chains is the Metropolis–Hastings algorithm, because it is not very restrictive with respect to the posterior distribution. The advantage of this algorithm lies in the fact that the samples are generated from a proposed auxiliary distribution, in order to facilitate the obtention of the sample values [74].

6.5 Results and Discussion

For the analysis and verification of the effectiveness of the methods of solution of the direct and inverse problem presented in this chapter, it was considered the hypothetical case of a channel with dimensions of $L_x = 200$ m length by $L_z = 1$ m depth, with a continuous and constant flow and substance concentration entering the domain at $x = 0$. For the longitudinal velocity $U(z)$ a logarithmic profile was plotted as a function of the friction velocity near the bottom, see Eq. (6.2).

The values of the longitudinal and vertical dispersion coefficients, D_x and D_z, respectively, and the other information used in the simulations are shown in Table 6.1.

The convergence behavior of the solution as a function of the series truncation order, N_{tr}, is shown in Tables 6.2, 6.3, 6.4 and 6.5.

It is observed, for most results shown in Tables 6.2, 6.3, 6.4 and 6.5, that the series converged to at least two significant digits, demonstrating that only five terms of the series are enough to obtain a good convergence of the model solution, i.e. $N_{tr} = 5$.

In all subsequent simulations presented in this chapter, five terms were used in the series expansion of the solution of the problem, i.e. $N_{tr} = 5$, which was sufficient to obtain the desired precision in at least two digits of convergence to the concentration field. In Fig. 6.1 it is shown the temporal evolution of sediment concentrations calculated with the GITT for the central region of the flow.

Table 6.1 Values of the parameters used in the simulations

Parameters	Value
Longitudinal dispersion coefficient, D_x [m^2 s^{-1}]	0.25
Vertical dispersion coefficient, D_z [m^2 s^{-1}]	0.006
Sediment settling velocity, W_s [m s^{-1}]	0.013
Reference concentration in the flow bed, C_a [mg L^{-1}]	100
Continuous emission concentration C_0 $[mgL^{-1}]$	50
Initial sediment concentration C_1 $[mgL^{-1}]$	0
L_x [m]; L_z [m]	200; 1
Launch position z [m], at $x = 0$ [m]	1
Erosion and deposition fluxes, $F_E = F_D$ [kg m^{-2} s^{-1}]	0.15
Mean velocity, $\overline{U}$ [m s^{-1}]	1
Friction velocity, u^* [m s^{-1}]	0.15

Table 6.2 Convergence analysis in relation to the order of truncation ($z = 0.5$ m, $t = 50$ s)

x [m]	$N_{tr} = 2$	$N_{tr} = 5$	$N_{tr} = 10$	$N_{tr} = 20$	$N_{tr} = 50$	$N_{tr} = 100$
0	47.939841	51.307159	49.558268	49.901118	50.024095	49.996225
40	40.157284	39.825171	39.967085	39.990508	39.991084	39.991274
80	40.018214	39.658897	39.797702	39.820180	39.820842	39.820954
120	40.018214	39.658597	39.797816	39.820180	39.820842	39.820954
160	40.018214	39.658591	39.797815	39.820180	39.820842	39.820954
200	40.018214	39.658591	39.797815	39.820180	39.820842	39.820954

Table 6.3 Convergence analysis in relation to the order of truncation ($z = 0.5$ m, $t = 100$ s)

x [m]	$N_{tr} = 2$	$N_{tr} = 5$	$N_{tr} = 10$	$N_{tr} = 20$	$N_{tr} = 50$	$N_{tr} = 100$
0	47.939841	51.307159	49.558268	49.901118	50.024095	49.996225
40	40.020650	40.032548	40.087023	40.099034	40.099979	40.100056
80	33.827856	33.783115	33.804730	33.809621	33.809840	33.809848
120	32.040868	32.003207	32.011488	32.013176	32.013237	32.013302
160	31.835282	31.815674	31.819277	31.819709	31.819718	31.819751
200	31.834974	31.815374	31.818993	31.819424	31.819429	31.819457

After 200 s of simulation, the permanent sediment transport regime is observed. In Fig. 6.2 it is shown the solution of the direct problem with both the GITT and the Mathematica's NDSolve routine for the concentration of non-cohesive sediments along the domain of interest.

In order to compare the two methods of solution, it is considered the dimensionless relative error given by the expression:

Table 6.4 Convergence analysis in relation to the order of truncation ($x = 100$ m, $t = 50$ s)

z [m]	$N_{tr} = 2$	$N_{tr} = 5$	$N_{tr} = 10$	$N_{tr} = 20$	$N_{tr} = 50$	$N_{tr} = 100$
0	103.22075	102.95371	103.14173	103.17629	103.17748	103.17779
0.2	70.386478	70.095591	70.251324	70.278406	70.279383	70.279618
0.4	48.294629	47.954539	48.089713	48.117588	48.118407	48.118562
0.6	33.123089	32.755102	32.890390	32.908253	32.908868	32.908941
0.8	22.496013	22.154797	22.237880	22.254420	22.254801	22.254818
1.0	14.897526	14.641833	14.678076	14.683198	14.683253	14.683225

Table 6.5 Convergence analysis in relation to the order of truncation ($x = 100$ m, $t = 100$ s)

z [m]	$N_{tr} = 2$	$N_{tr} = 5$	$N_{tr} = 10$	$N_{tr} = 20$	$N_{tr} = 50$	$N_{tr} = 100$
0	97.086383	96.901276	96.966154	96.977399	96.977787	96.977803
0.2	64.410352	64.220646	64.269194	64.276983	64.277280	64.277292
0.4	42.780342	42.599079	42.641526	42.650652	42.650923	42.650958
0.6	28.338870	28.207303	28.253591	28.258879	28.259106	28.259157
0.8	18.660722	18.599136	18.623001	18.628463	18.628627	18.628678
1	12.176320	12.152168	12.159044	12.160152	12.160213	12.160242

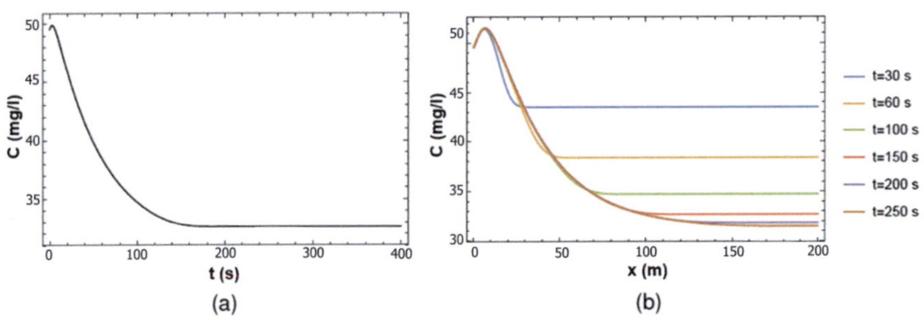

Fig. 6.1 Temporal distribution of sediment concentration: (**a**) $x = 100$ m and $z = 0.5$ m; (**b**) $z = 0.5$ m, for different time instants

$$E_r(x, z, t) = \frac{C_{GITT}(x, z, t) - C_{NDSOLVE}(x, z, t)}{C_a} \times 100\% \tag{6.14}$$

The spatial distribution of the suspended sediment concentration in steady state, i.e. after 200 s of simulation, is presented in Fig. 6.3.

The results obtained presented a good agreement with an error of up to 4% with respect to the background reference concentration. The solutions by GITT and the NDSolve routine had a computational cost of 0.30 s and 11.89 s, respectively using a computer with Intel Core i5 processor, 2.67 GHz and 4 GB of RAM. With GITT, the computational gain was high, generally being on the order of 40 times faster than

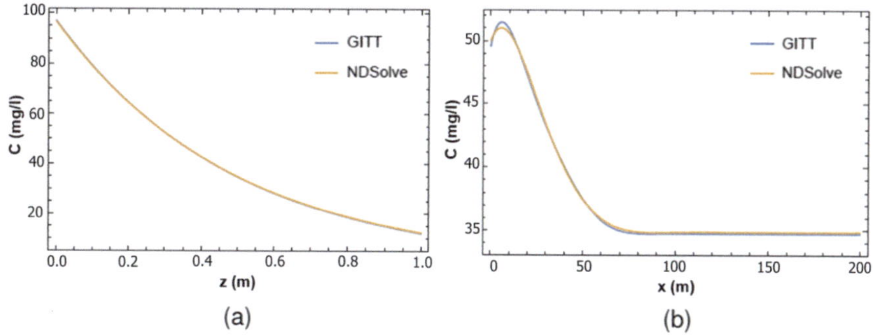

Fig. 6.2 Vertical and longitudinal distribution of sediment concentration at $x = 100$ m (**a**), and z = 0.5 m (**b**), respectively, both at time $t = 200$ s

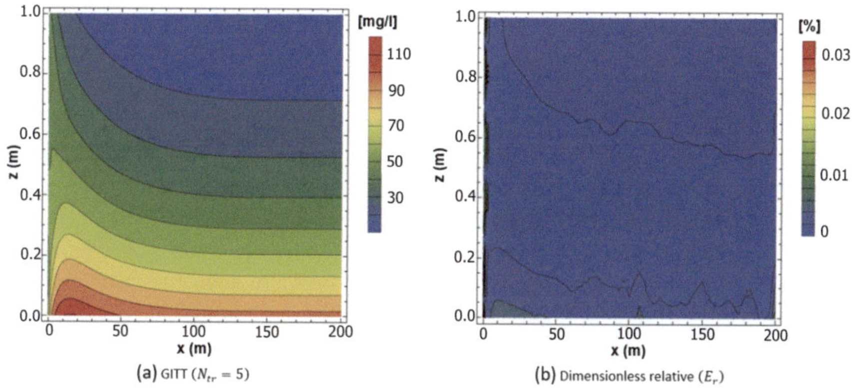

Fig. 6.3 Spatial distribution of sediment concentration at $t = 200$ s (**a**); and dimensionless relative error at the same time instant

the NDSolve Routine, demonstrating the adequacy of the use of this technique for the analysis of the subsequent inverse problems of interest presented in this chapter.

The first step in the solution of the inverse problem was to perform a sensitivity analysis to study the effect that the variation of the dispersive and transport parameters can cause in the results. This analysis contributes significantly to the formulation and solution of inverse problems, since it allows, among other things, the choice of experimental data that results in better estimates for the parameters of interest [15].

In this work, due to the complexity of the physical problem, and the lack of knowledge of an analytical solution for the direct problem, the modified (scaled) sensitivity coefficients were calculated numerically using a centered finite difference approach. For this approximation the following expression was considered:

$$X_{P_j}(x,z,t) = P_j \times \left(\frac{C(x,z,t,P_j+\varepsilon_j) - C(x,z,t,P_j-\varepsilon_j)}{2\varepsilon_j} \right), j = 1,2,\ldots,N_p \tag{6.15}$$

where N_p is the total number of unknown parameters to be estimated, and each sensitivity coefficient X_{P_j} is a measure of the sensitivity of the sediment concentration C calculated with respect to the variations ε_j occurring in the parameter P_j.

In order to obtain good estimates for the parameters of interest it is important that such parameters lead to high sensitivity coefficients. This means that small variations in the respective parameters cause significant changes in the response of the model. In Figs. 6.4 and 6.5, the curves for each sensitivity coefficient, obtained numerically with centered finite difference approximations, with perturbations$\varepsilon_1 = 10^{-1}$ ms^{-1}, $\varepsilon_2 = 10^{-3}$ m^{2}s^{-1}, $\varepsilon_3 = 10^{-1}$ m^{2}s^{-1} and $\varepsilon_4 = 10^{-2}$ m s^{-1} for the unknown parameters D_x, D_z, u^* and W_s, respectively.

The obtained graphs show the transient behavior of the sensitivity coefficients for the four parameters involved in the inverse analysis, considering sensors located in different spatial positions, $x = 30$, 60, 100 and 150 m, with z fixed at 0.5 m, and $z = 0.2$, 0.4, 0.6 and 0.8 m, with x fixed at 100 m.

The largest variations for most sensitivity coefficients occurred in the first 150 s. It is observed that the coefficients of sensitivity for the parameters D_x and u^* presented relatively small values in comparison to the other coefficients.

Low sensitivity values may cause an inability to identify parameters in the inverse analysis, which can be explained by the fact that in many advective problems the phenomena investigated dominates in the x-direction, and longitudinal dispersion is commonly neglected [97].

Due to the difficulty in estimating the parameters D_x and u^*, a priori information was considered for these two parameters, according to a normal distribution with mean values of, 0.25 m^{2}s^{-1} and 0.15 ms^{-1}, respectively, and standard deviation of 20% of each mean value.

As there were no real experimental data available for the inverse analysis, synthetic data were obtained at the position $x = 100$ m and $z = 0.5$ m where the transient values of the sensitivity coefficients were predominantly higher, and thus becoming the preferable site for collecting the experimental information. To the calculated values of the concentration using the exact values of the parameters of the vector of unknowns P_{exact} were added random noise according to a normal distribution with zero mean and known standard deviation σ_{exp}, as follows:

$$C_{exp_i} = C_{calc_i}(P_{exact}) + r_i, \quad r_i \sim \zeta(0, \sigma_{exp}), \quad i = 1,2,\ldots,N_d \tag{6.16}$$

The synthetic experimental data were generated with two different values for σ_{exp} in Eq. (6.16), 2.5 and 5.0 mgL^{-1} for a total of 150s of simulation of the direct problem. In Eq. (6.16) ζ represents the normal distribution.

The results for the inverse problem obtained with the Monte Carlo method via Markov Chains are presented in Figs. 6.6 and 6.7. The parameters of interest were

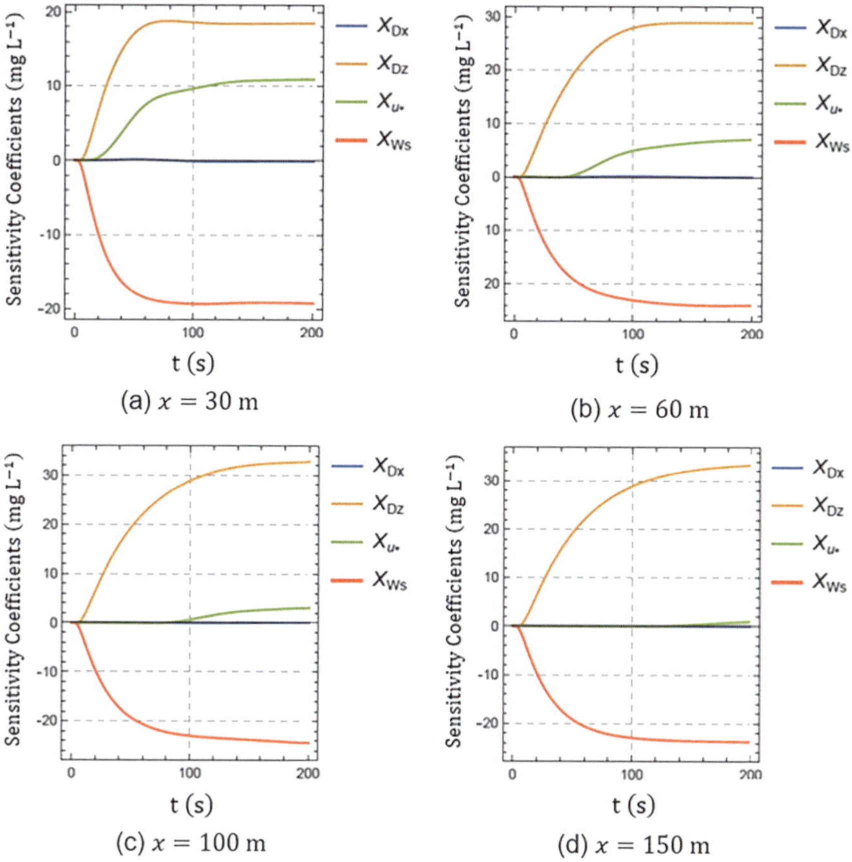

Fig. 6.4 Modified (scaled) sensitivity coefficients for the parameters D_x, D_z, u^*_* and W_s at the position $z = 0.5$ m

estimated using the Metropolis–Hastings algorithm, accepting or rejecting at each iteration the randomly generated sample as the inverse problem solution candidate. It is observed from the figures that the convergence of the chains obtained by the MCMC occurs even at the initial stages of the iterative process. The quantity N_{MCMC} in the horizontal axes of Figs. 6.6 and 6.7 represents the number of states in the MCMC method.

Considering different noise levels, the MCMC proved to be efficient, and was able to reach the equilibrium state for all analyzed parameters.

It was considered a burn-in period of 10,000 states in the Markov chain for a total of 50,000 samples throughout the chain. Taking the sample mean of the remaining 40,000 states, we obtained the estimates for the parameters of interest. The corresponding histograms, presented in Fig. 6.8, the solution considering the smaller level of noise, that is, using experimental data with $\sigma_{exp} = 2.5\,\text{mg L}^{-1}$, the parameters distribution seems to be well represented using the normal distribution. In Fig. 6.9,

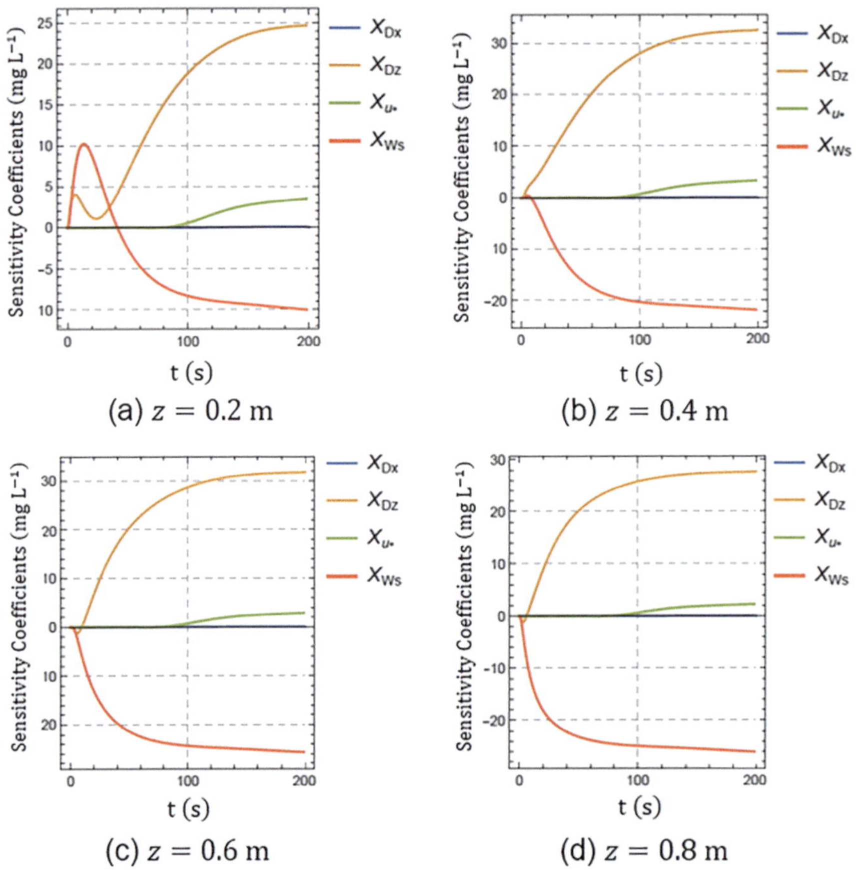

Fig. 6.5 Modified (scaled) sensitivity coefficients for the parameters D_x, D_z, u^* and W_s at the position $x = 100$ m

representing the histograms for the case considering the higher noise level, that is, using experimental data with $\sigma_{exp} = 5.0\,\text{mg L}^{-1}$, the non-linear effect of the inverse problem seems to be more important in the error propagation, and the parameters distribution seem to present the characteristic of a skewed normal distribution.

In order to facilitate the comparison between the results found with the MCMC, in Tables 6.6 and 6.7 are presented the main statistical properties of the parameters of interest with mean values, standard deviation, coefficients of variation, Root Mean Square Error (RMSE), and credibility intervals with 95% confidence.

By analyzing the data from Tables 6.6 and 6.7, it is verified that all the parameters exact values are within the respective credibility intervals. In addition, a relatively low RMSE value was obtained.

Out of the 40,000 Markov Chain samples generated as solution candidates, 12,568 (31.42%) and 10,192 (25.48%) states, for the inverse problems solutions with $\sigma_{exp} =$

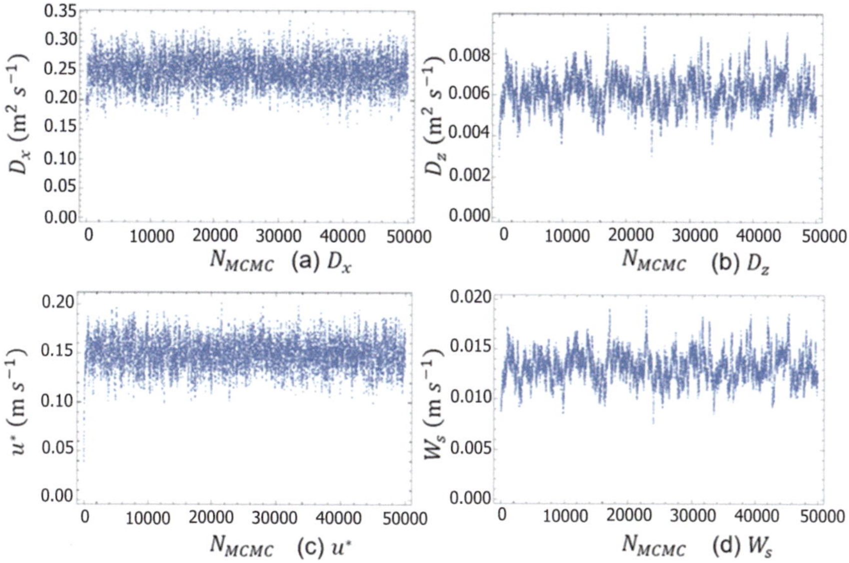

Fig. 6.6 Evolution of the Markov Chains for the solution of the inverse problem considering experimental data with $\sigma_{exp} = 2.5\,\text{mg L}^{-1}$

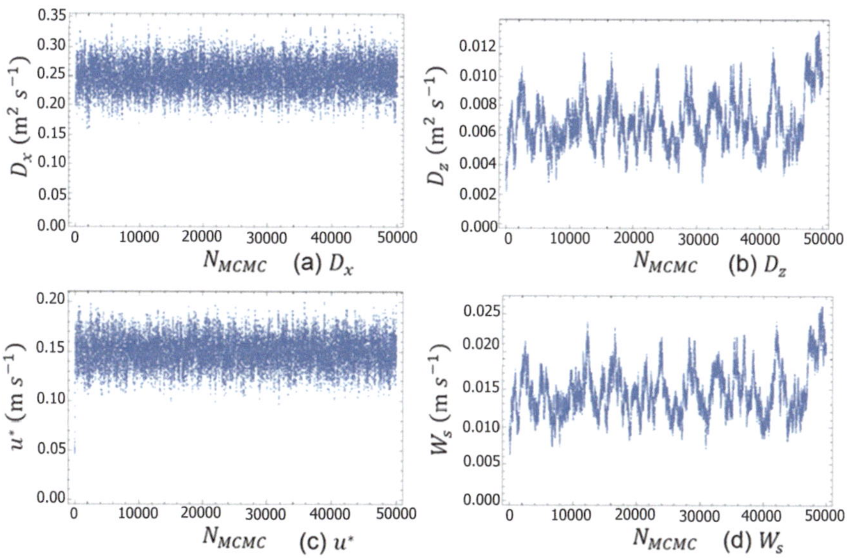

Fig. 6.7 Evolution of the Markov Chains for the solution of the inverse problem considering experimental data with $\sigma_{exp} = 5.0\,\text{mg L}^{-1}$

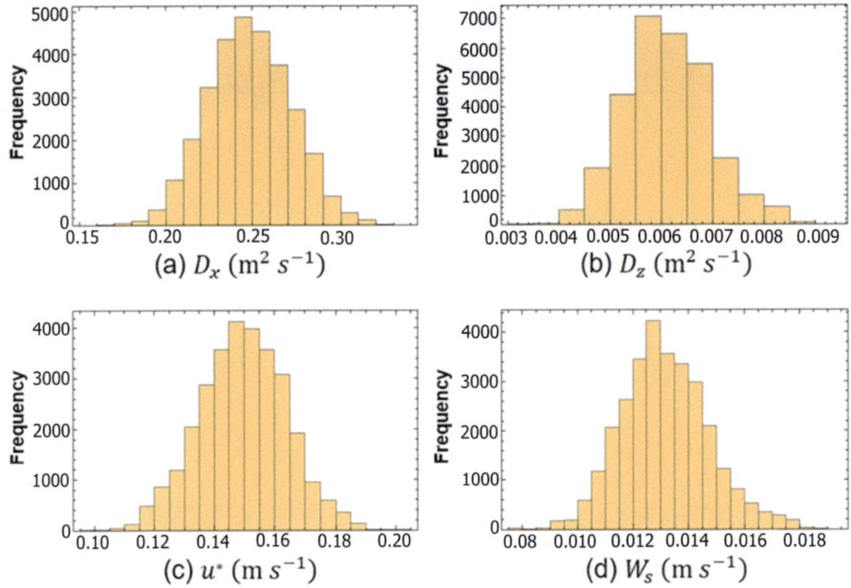

Fig. 6.8 Histograms for the solution of the inverse problem considering experimental data with $\sigma_{exp} = 2.5\,\text{mg L}^{-1}$

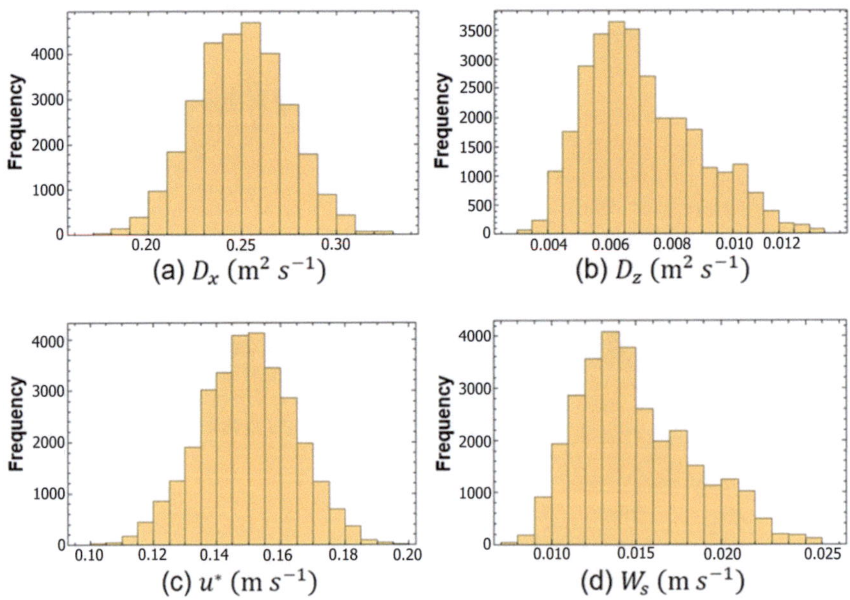

Fig. 6.9 Histograms for the solution of the inverse problem considering experimental data with $\sigma_{exp} = 5.0\,\text{mg L}^{-1}$

Table 6.6 Main statistical properties of the estimated parameters obtained with the solution of the inverse problem considering experimental data with $\sigma_{exp} = 2.5mgL^{-1}$. See Eq. (6.16)

Parameter	Exact value	$\bar{\mu}$	σ	$\frac{\sigma}{\bar{\mu}}[\%]$	C.I. (95%)
$D_x[m^2s^{-1}]$	0.25	0.249819	0.024735	10.346	[0.201042;0.297235]
$D_z[m^2s^{-1}]$	0.006	0.005912	0.000614	10.385	[0.004796;0.007128]
u^* $[ms^{-1}]$	0.15	0.154260	0.012655	8.203	[0.129280;0.177654]
$W_s[ms^{-1}]$	0.013	0.012806	0.00116	9.058	[0.010677;0.015115]
Acceptance rate (%)	31.42				
RMSE (mgL^{-1})	0.00810335				

Table 6.7 Main statistical properties of the estimated parameters obtained with the solution of the inverse problem considering experimental data with $\sigma_{exp} = 5.0mgL^{-1}$. See Eq. (6.16)

Parameter	Exact value	$\bar{\mu}$	σ	$\frac{\sigma}{\bar{\mu}}[\%]$	C.I. (95%)
$D_x[m^2s^{-1}]$	0.25	0.249987	0.025182	10.073	[0.200224;0.299543]
$D_z[m^2s^{-1}]$	0.006	0.006265	0.001195	19.080	[0.004272;0.008894]
$u^*[ms^{-1}]$	0.15	0.149579	0.014906	9.965	[0.119895;0.179262]
$W_s[ms^{-1}]$	0.013	0.013447	0.002223	16.531	[0.009765;0.018380]
Acceptance rate (%)	25.48				
RMSE (mgL^{-1})	0.0131332				

2.5 and 5.0 mg L^{-1}, respectively, were considered in the statistical analysis. These percentual values correspond, therefore to the acceptance rates. The efficacy of the MCMC, and the good agreement between the exact and calculated parameters are demonstrated by the low RMSE values obtained.

Considering the mean values of the parameters obtained with the MCMC and the synthetic experimental data, in Fig. 6.10 it is shown the good quality of the inverse problem solution.

In general, a good agreement between the estimated and exact values of the dispersive and transport properties is observed, with a very small variation, demonstrating that the overall good fit between the experimental and theoretical curves is consistent.

As the estimates for the parameters involved in the model are accurate, it is possible to describe the sediment concentration profile, as well as the spatial distribution and temporal evolution of the concentrations in the entire domain of interest. Considering the mean values of the estimated parameters, and an experimental noise with standard deviation $\sigma_{exp} = 2.5$ mg L^{-1}, the spatial distribution of suspended sediment concentration in steady state after 200 s of simulation is presented in Fig. 6.11.

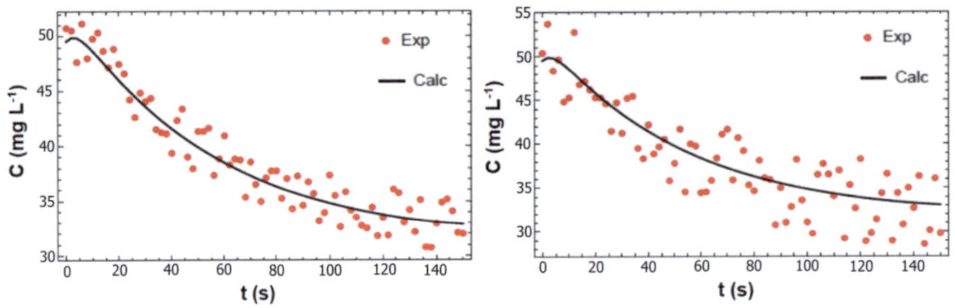

(a) $\sigma_{exp} = 2.5\ \mathrm{mg\ L^{-1}}$ (b) $\sigma_{exp} = 5.0\ \mathrm{mg\ L^{-1}}$

Fig. 6.10 Comparison between the sediment concentration calculated with the parameters obtained with the solution of the inverse problem, and the experimental data acquired at $x = 100$ m and $z = 0.5$ m

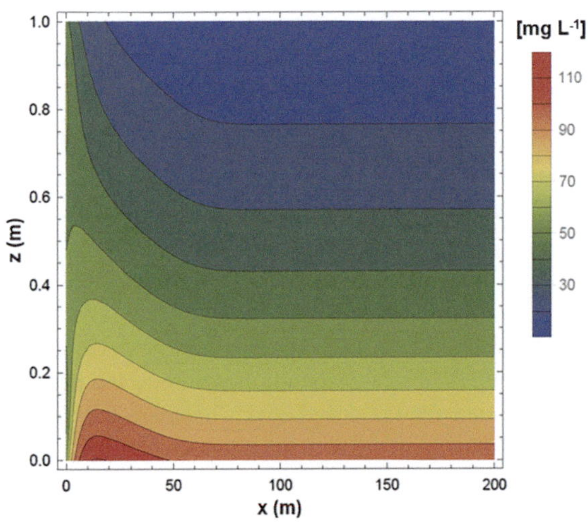

Fig. 6.11 Spatial distribution of sediment concentration at $t = 200$ s obtained with the solution of the direct problem using the estimated parameters obtained with the solution of the inverse problem and experimental data with $\sigma_{exp} = 2.5\ \mathrm{mg\ L^{-1}}$

6.6 Remarks

The methodology described in this chapter allowed, through the formulation and solution of the inverse problem, to estimate hydrodynamic and transport parameters, which are in general, difficult to measure, consequently making possible the observation of expected behaviors for the concentration of non-cohesive sediments in natural water bodies.

The use of the GITT technique in the solution of the direct problem of interest proved plausible and fast computationally, as it is desirable for the viability of the solution of the inverse problem. The sensitivity analysis allowed to interpret all the results obtained in a simple way, and to develop efficient experimental tools to obtain accurate estimates for the parameters of interest. The Markov Chain Monte Carlo

Method proved to be effective in solving the problem of identifying dispersive and transport parameters with an average execution time of approximately 9 h using a computer with Intel Core i5 processor, 2.67 GHz and 4 GB of RAM.

The use of hybrid analytical–numerical strategies through the GITT technique demonstrated a great potential for use in the solution of inverse problems with several possibilities of application. Obviously, for models with greater complexity, the use of the technique presents greater difficulty, but, when properly implemented, provides good approximations to the solution of the problem.

For future work, it is recommended to apply the MCMC to estimate the longitudinal velocity present in the advective term, as well as to set smooth functions that describe the fluxes present in the boundary conditions, especially on the surface and the bottom of the flow. Another important recommendation is the application of methodologies that allow the reduction of the computational cost of the inverse problem.

Chapter 7
Final Remarks

In this book, Chaps. 2–6 build up from the basic concepts presented in Chap. 1. In this chapter some final remarks are then presented, drawn from the research on pollutants transport in rivers and estuaries performed by the authors for over 15 years, pointing out also future trends that are to be faced, first in the model formulation and solution of Direct Problems, and finally in the Inverse Problems techniques that can be used to estimate parameters and functions of interest in environmental studies.

7.1 Fluvial Environment

In Chap. 2 several computer simulations are presented, studying hypothetical scenarios for the injection of the pollutant load into rivers, evaluating a real situation, the instantaneous launch in the low Macaé river. The simulations were carried out using the Finite Volume Method (FVM), adopting different meshes to minimize the effect of the truncation error and also the variation of the dispersive and advective terms, in order to check the best numerical adjustment considering experimental data collected in the field work. The results were obtained considering an instantaneous injection of the tracer in the center of the river, and another one close to its margin, to check the best point to discharge pollutants in streams. The formulation and modeling were able to satisfactorily represent pollutant dispersion phenomena, modeling the concentration peak with acceptable precision.

In Chap. 3 a numerical-analytical methodology, based on integral transforms, is used to solve the transport and the dispersion of a conservative substance in a river. The one-dimensional advection-transient solution with the Generalized Integral Transform Technique (GITT) considered a system truncated to fifty terms that was enough for the required accuracy.

J. Lugon Junior et al., *Inverse Problems in the Environmental Modelling of Rivers and Estuaries*, Studies in Systems, Decision and Control 246,
https://doi.org/10.1007/978-3-032-09937-2_7

7.2 Estuarine Environment

In Chap. 4 the Navier–Stokes equations generalized to a geophysical fluid is used to build the hydrodynamic and transport models using the MOHID Water platform. The models used the Finite Volume Method (FVM) to simulate the estuary of the Macaé River.

In Chap. 5, the same estuarine region of the Macaé River is considered to build the transport model with the Finite Element Method (FEM) for the discretization in the space and the Finite Differences Method (FDM) for discretization in time. The equations and methods are presented in detail, afterwards mesh variations are studied analyzing the consistency, stability and convergence issues of the discretization method.

In Chap. 6, the GITT technique is used to solve the hydrodynamic and transport models to simulate the concentration of non-cohesive sediments in natural water bodies.

7.3 Environmental Modelling Inverse Problems

In Chap. 2, after performing a sensitivity analysis, both transverse and longitudinal dispersion coefficients are calibrated using experimental data collected in an experimental work carried out in the low Macaé River region. The Finite Volume Method (FVM) was used to solve the Direct Problem and the simulated results were used to train an Artificial Neural Network (ANN). The discrepancies between the values observed and calculated after the peak concentration motivates future investigation. A trial-and-error approach is used in the calibration of the model.

In Chap. 3, the main objective is the estimation of a function representing one of the boundary conditions of the one-dimensional advection–dispersion equation in a transient regime. The Inverse Problem is solved using the Markov Chain Monte Carlo Method with two non-informative priors. The proposed methodology is able to satisfactorily recover the way a pollutant enters the modeled water medium, even with the low levels of information used. Since the 80,000 samples (states) defined in the MCMC were not sufficient for them to achieve equilibrium, due to the high computational cost, future investigation with a larger number of states is to be performed.

In Chap. 4, the calibration of hydrodynamic models with different stochastic techniques is performed: GA, PCA, LJ, PSO and ANN. This is a very important topic to research on, because the best technique to be used depends on the model and even on the specific case of study.

In Chap. 5, the case of study is the estimation of the location and intensity of a hypothetical continuous punctual discharge in the estuarine region of the Macaé River. Using the sensitivity analysis, it is possible to observe that these parameters are correlated, which makes it difficult to apply inverse problem techniques. Different

Inverse Problems solution methods and combinations are tested, with good results to the estimation of the position and intensity of the source. Further experimental studies are to be investigated in the future for practical environmental applications.

In Chap. 6, the estimation of hydrodynamic and transport parameters, that are, in general, difficult to measure, aim at the study of the concentration distribution of non-cohesive sediments in natural water bodies. The Markov Chain Monte Carlo Method proves to be effective in solving the problem of identifying dispersive and transport parameters. For the future remains to be investigated the use of the MCMC to estimate the longitudinal velocity present in the advective term, as well as to set smooth functions that describe the fluxes present in the boundary conditions, especially on the surface and the bottom of the flow.

References

1. Abbot, M.B., Basco, D.R.: Computational Fluid Dynamics, an Introduction for Engineers. [S.l.]: Logan Group, UK Limited (1989)
2. Abbott, M.B., Damsgaardand, A., Rodenhuis, G.S.: System 21, Jupiter, a design system for two dimensional nearly horizontal flows. J. Hydraul. Res. **11**(1), 1–28 (1973)
3. Almoaia, A.E.N.F., Sacco, W.F., Silva Neto, A.J.: Hybrid differential evolution with the topographical heuristic. Learn. Nonlinear Models **17**(2), 42–52 (2019)
4. Alvarez Acevedo, N.I., Roberty, N.C., Silva Neto, A.J.: A one-dimensional inverse radiative transfer problem with time-varying boundary conditions. Inverse Probl. Sci. Eng. **12**(2), 123–140 (2004)
5. Alvarez Acevedo, N.I., Roberty, N.C., Silva neto, A.J.: An explicit formulation for the inverse transport problem using only external detectors—Part I: Computational modelling. Comput. Appl. Mathem. **29**(3), 343–358 (2010)
6. Alvarez Acevedo, N.I., Roberty, N.C., Silva Neto, A.J.: An explicit formulation for the inverse transport problem using only external detectors—Part II: application to one-dimensional homogeneous and heterogeneous participating media. Comput. Appl. Mathem. **29**(3), 359–374 (2010)
7. Amaral, K.J.: Macaé River Estuary: Computational Modeling as a Tool for Integrated Water Resources Management. Master Dissertation. UFRJ. Rio de Janeiro (2003). (in Portuguese)
8. ANA (Brazilian National Water Agency). Technical Manual on Measurement of Liquid Discharge in Great Rivers. 2o. ed. Brasília: [s.n.], 2014. 94 p. E-book. Water Resources in Brazil: Report 2016. Brasília: [s.n.] (2016). 95 p. E-book. (in Portuguese)
9. Angelini, R.: Ecosystems and ecological modeling. In: Pompêo, M.L.M. (Ed.) Perspectives of Limnology in Brazil. São Luís: Gráfica e Editora União, 198 p. (1999). (in Portuguese)
10. Arakawa, A., Lamb, V.R.: Computational design of the basic dynamical processes of the UCLA General Circulation Model. Methods Comput. Phys. **17**, 174–264 (1977)
11. Assad, L.P.F., Mano, M.F., Decco, H.T., Torres Junior, A.R., Landau, L.: Basic Notion of Hydrodynamic Computational Modelling and Pollutant Dispersion, Ed. COPPE / Rio de Janeiro Federal University, Rio de Janeiro: COPPE/UFRJ (2009). (in Portuguese)
12. Barbosa Jr. A.R., Silva, G.Q., Devens, J.A., Neves, B.V.B.: Direct methods of determining the coefficient of longitudinal dispersion in natural watercourses. Part 2: Application and comparison of methods. Revista da Escola de Minas, Ouro Preto, v. 58, n.2, p. 209–214, jul./set. (2005). (in Portuguese).
13. Barros, F.P.J.D., Mills, W.B., Cotta, R.M.: Integral transform solution of a two-dimensional model for contaminant dispersion in rivers and channels with spatially variable coefficients. Environ. Model. Softw. **21**(5), 699–709 (2006)

J. Lugon Junior et al., *Inverse Problems in the Environmental Modelling of Rivers and Estuaries*, Studies in Systems, Decision and Control 246,
https://doi.org/10.1007/978-3-032-09937-2

14. Bates, P.D., Lane, S.N., Ferguson, R.I.: Computational fluid dynamics: application in environmental hydraulics. [S.l.]: WILEY, 2005. BEQUETTE, F. El Agua: ¿Una Crisis Inminente? Revista El Correo de la UNESCO (1998)
15. Beck, J.V., Blackwell, B.C., St. Clair Jr., C.R.: Inverse Heat Conduction: III Posed Problems. New York: Wiley Interscience (1985)
16. Bequette, F.: El Agua: ¿Una Crisis Inminente? Revista El Correo de la UNESCO (1998)
17. Bird, R.B., Stewart, W.E., Lightfoot, E.N.: Transport Phenomena. Wiley, New York (1960)
18. Bishop, C.M.: Neural Network for Pattern Recognition. Oxford University Press. inc., New York (1995)
19. Boretti, A., Rosa, L: Reassessing the Projections of the World Water Development Report, npj Clean Water **2**, 15.1–6 (2019)
20. Bortoli, A.L. de: Introduction to Computational Fluid Dynamics. Porto Alegre: Universidade Federal do Rio Grande do Sul (2000). (in Portuguese)
21. Braga, B., Hespanhol, B., Conejo, J.G.L., Barros, M.T.L., Spencer, M., Porto, M., Nucci, N., Juliano, N., Eiger, S.: Introduction to Environmental Engineering. 2. ed. São Paulo: Pearson (2005). (in Portuguese)
22. Brazilian Navy, Nautic Charts, https://www.marinha.mil.br/chm/dados-do-segnav/cartas-raster. Accessed April 2025
23. Brazilian navy. Brazilian Maregraphic Stations, 2023. Available from: https://www.marinha.mil.br/chm/estacoes-maregraficas. Accessed April 2025. (in Portuguese).
24. Buske, D., Vilhena, M.T., Moreira, D.M., Tirabassi, T.: Simulation of pollutant dispersion for low wind conditions in stable and convective planetary boundary layer. Atmospheric Environ. Elsevier **41**(26), 5496–5501 (2007)
25. Camps Echeverría, L., Llanes Santiago, O., Campos Velho, H.F., Silva Neto, A.J.: Fault Diagnosis Inverse Problems: Solution with Metaheuristics, Springer (2019)
26. Cassol, M., Wortmann, S., Rizza, U.: Analytic modeling of two-dimensional transient atmospheric pollutant dispersion by double GITT and Laplace Transform techniques. Environ. Model. Software Elsevier **24**(1), 144–151 (2009)
27. Castellões, F.V., Cardoso, C.R., Couto, P., Cotta, R.M.: Transient analysis of slip flow and heat transfer in microchannels. Heat Trans. Eng. Taylor & Francis **28**(6), 549–558 (2007)
28. Castellões, F.V., Cotta, R.M.: Analysis of transient and periodic convection in microchannels via integral transforms. Progress Comput. Fluid Dyn. Int. J. Inderscience Publ. **6**(6), 321–326 (2006)
29. CETESB (São Paulo State Environmental Company). Fundamentals of Water Pollution Control. Available from: https://cetesb.sp.gov.br/posgraduacao/wp-content/uploads/sites/33/2018/11/Apostila-Fundamentos-do-Controle-de-Polui%C3%A7%C3%A3o-das-%C3%81guas-T3.pdf. Accessed in April, 2025. 2018. (in Portuguese)
30. Chapra, S.C.: Surface Water—Quality Modeling. Waveland Press, Inc, Long Grove, Illinois (200)
31. Chippada, S., Dawson, C., Wheeler, M.: Agodonov-type finite volume method for the system of shallow water equations. Comput. Methods Appl. Mech. Eng. **151**(01), 105–130 (1998)
32. Clift, R., Grace, J.R., Weber, M.E.: Bubbles, drops and particles. J. Fluid Mech.. Academic Press, New York **94**(4), 795–796 (1978)
33. Conama (National Council for the Environment), Resolution number 357/2005. Provides for the classification of water bodies and environmental guidelines for their classification, as well as establishing the conditions and standards for effluent discharge, and provides other measures. DOU Brasília, DF, n. 053, p. 58–63 (2005). (in Portuguese)
34. Conama (National Council for the Environment), Resolution number 430/2011. Provides for conditions and standards for the discharge of effluents, complements and alters Resolution No. 357/2005, from the National Council for the Environment—CONAMA. (in Portuguese).
35. Costa, H.: Subsidies for water resources management in the Macacu, São João, Macaé and Macabu river basins. Rio de Janeiro: Secretary of State for the Environment and Sustainable Development (SEMADS) (1999). (in Portuguese).

36. Costa, D.A., Bayissa, Y., Barbosa, K.V., Villas-Boas, M.D., Bawa, A., Lugon Jr., J., Silva Neto, A.J., Srinivasan, R.: Water quality estimates using machine learning techniques in an experimental watershed. J. Hydroinformatics **26**(11), 2798–2814 (2024)
37. Costa, D.A., Bayissa, Y., Villas-Boas, M.D., Maskey, S., Lugon Jr., J., Silva Neto, A.J., Srinivasan, R.: Water Availability and Extreme Events Under Climate Change Scenarios in an Experimental Watershed of the Brazilian Atlantic Forest, Science of the Total Environment, Vol. 946, pp. 174417.1–15 (2024)
38. Cotta, R.M.: Integral Transforms in Computational Heat and Fluid Flow. CRC Press (1993)
39. Cotta, R.M., Gerk, J.E.V.: Mixed finite-difference/integral transform approach for parabolic-hyperbolic problems in transient forced convection. Numer. Heat Transfer Part B Fund. Taylor & Francis **25**(4), 433–448 (1994)
40. Cotta, R.M., Mikhailov, M.D.: Heat Conduction: Lumped Analysis, Integral Transforms, Symbolic Computation. Wiley Chichester, Chichester (1997)
41. Derrac, J., García, S., Molina, D., Herrera, F.: A practical tutorial on the use of nonparametric statistical tests as a methodology for comparing evolutionary and swarm intelligence algorithms. Swarm Evol. Comput. **1**(1), 3–18 (2011)
42. Devens, J.A.: Quantification of the longitudinal dispersion coefficient in small natural water-courses using an environmentally neutral tracer. Master Dissertation—Universidade Federal de Ouro Preto, Ouro Preto (2006). (in Portuguese)
43. Domínguez, D.E.M., Hernández, O.R.: A numerical model of contaminant transport in channels applying the finite element method. Ingeniería Hidráulica y Ambiental. **28**(3) (2007). (in Spanish)
44. Dowding, K.J., Blackwell, B.F.: Applications of sensitivity coefficients for heat. Numer. Heat Transfer **36**, 33–55 (1999)
45. Dyer, K.: Estuaries: A Physical Introduction. [S.l.]: John Wiley & Sons (1997)
46. Eiger, S.: Water quality in rivers and estuaries. In: Branco, S M., Porto, R.L.L., Cleary, R.W. (Ed.). Environmental Hydrology. São Paulo: EDUSP/ABRH, 1991, Water Resources ABRH Colection, v. 3 (in Portuguese)
47. Elder, J.W.: The dispersion of marked fluid in turbulent shear flow. J. Fluid Mech. **5**, 544–560 (1959)
48. Engelbrecht, A.P.: Computational Intelligence: An Introduction. 2.ed. [S.l.]: John Wiley & Sons (2007)
49. Estrada, V., Parodi, E.R., Diaz, M.S.: Determination of biogeochemical parameters in eutrophication models with simultaneous dynamic optimization approaches. Comput. Chem. Eng. **33**, 1760–1769 (2009)
50. Etemad-Shahidi, A., Imberger, J.: Optimal estimation of parameters for a estuarine eutrophication model. Dyn. Atmos. Oceans **39**(3–4), 175–187 (2004)
51. Fan, F.M., Fleischmann, A.S., Collischonn, W., Ames, D.P., Rigo, D.: Large-scale analytical water quality model coupled with GIS for simulation of point sourced pollutant discharges. Environ. Model. Software Elsevier **64**, 58–71 (2015)
52. Fernandes, R.M.A.S.: Operational Modeling in the Tagus Estuary. Master Dissertação—Instituto Superior Técnico, Universidade Técnica de Lisboa, Lisboa (2005). (in Portuguese)
53. Fischer, H.B., Imberger, J., List, E.J., Koh, R.C.Y., Brooks, N.H.: Mixing in Inland and Coastal Waters. Academic Press, New York (1979)
54. Fletcher, C.A.J.: Computational Techniques for Fluid Dynamics. Volume I. 2nd Edition. Springer Series in Computational Physics. New York: Springer Verlag. p. 401 (1991)
55. Fonseca, H.M.: Solutions of Inverse Problems with Thermographic Measurements in Micro-Scales. 2012. 142 p. Thesis (Mechanical Engineering Doctor) Universidade Federal do Rio de Janeiro, Rio de Janeiro (2012). (in Portuguese)
56. Fortuna, A.O.: Computational Techniques for Fluid Dynamics. São Paulo: EdUSP (2000). (in Portuguese)
57. Friguis, M.S., Knupp, D.C., Abreu, L.A.S., Silva Neto, A.J.: An Inverse diffusive population problem with the maximum a posteriori Bayesian approach and the Luus-Jaakola method. 2nd International Symposium of Modeling Applied to Engineering (MAI2016), 18 Convención Científica de Ingeniería y Arquitectura, Havana, Cuba (2016)

58. Gallo Neto, H., Camba, D., Jeveaux, J.L., Barbosa, C.B.M.V., Albaladejo, M.C., Lugon Jr., J., Kalas, F.A., Rodrigues, P.P.G.W., Silva Neto, A.J.: Carcasses of Cetaceans Found in the North Coast of São Paulo: Management Strategies and Computational Modeling, IX Congresso sobre Planejamento e Gestão das Zonas Costeiras dos Países de Expressão Portuguesa, Lisboa, Portugal (2019)
59. Galski, R.L., De Sousa, F.L., Ramos, F.M., Silva neto, A.J.: Application of a GEO+SA hybrid optimization algorithm to the solution of an inverse radiative transfer problem. Inverse Probl. Sci. Eng. **17**(3), 321–334 (2009)
60. Gamerman, D., Lopes, H.F.: Markov Chain Monte Carlo: Stochastic Simulation for Bayesian Inference. [S.l.]: CRC Press, 2nd Ed. (2006)
61. García, S., Molina, D., Lozano, M., Herrera, F.: A Study on the use of non- parametric tests for analyzing the evolutionary algorithms' behaviour: a case study on the CEC'2005 special session on real parameter optimization. J. Heuristics **15**(6), 617–644 (2009)
62. Gill, A.: Atmosphere-Ocean Dynamics. International Geophysics Series, Vol. 30, Academic Press, San Diego (1989)
63. Graham, C.: Markov Chains: Analytic and Monte Carlo Computations. Wiley (2014)
64. Guerrero, J.S.P., Pontedeiro, E., Genuchten, M.T. van, Skaggs, T.: Analytical solutions of the one-dimensional advection–dispersion solute transport equation subject to time-dependent boundary conditions. Chem. Eng. J. **221**, 487–491 (2013). Elsevier
65. Guimarães, M.T.G., Abreu, L.A.S., Knupp, D.C., Silva Neto, A.J.: An Explicit Formulation for the Thermal Contact Conductance Estimation Using Truncated Eigenfucntion Expansions Regularization, 24th International Congress of Mechanical Engineering, ABCM, Curitiba, Brasil (2017)
66. Harte, J.: Consider a Cylindrical Cow: More Adventures in Environmental Problem Solving. Sausalito, California: University Science Books, 211 p. (2001)
67. Haykin, S.: Neural Networks: A Comprehensive Foundation. Perason Prentice Hall, Singapore (1999)
68. Holland, J.H.: Adaptation in Natural and Artificial Systems. [S.l.]: Ann Arbor, MI: University of Michigan Press (1975)
69. Holley, E.R.: United view of diffusion and dispersion. J. Hydraulics Div. **95**, 621 (1969). Proceeding of the American Society of Civil Engineers
70. Horritt, M.S., Bates, P.D., Mattinson, M.J.: Effects of mesh resolution and topographic representation in 2D finite volume models of shallow water fluvial flow. J. Hydrol. **329**, 306–314 (2006)
71. Hsu, M.-H., Kuo, A. Y., Kuo, J.-T., Liu, W.-C.: Procedure to calibrate and verify numerical models of estuarine hydrodynamics. J. Hydraulic Eng., 166–182 (1999)
72. Hughes, T.J.R.: The Finite Element Method: linear static and dynamic finite element analysis, Englewood Cliffs, N.J.: Prentice-Hall, 676 p. (1987)
73. INAG: Economic analysis of water uses in the context of the Water Framework Directive. Water Economics Group (GEA)—Direcção de Serviços de Planeamento (DSP) do Instituto da Água (INAG), Lisboa (2001). (in Portuguese)
74. Kaipio, J.P., Somersalo, E.: Computational and statistical methods for inverse problems. Applied mathematical sciences, Springer Berlin, v. 160 (2004)
75. Kanji, G.K.: 100 Statistical Tests. 3rd ed. Sage Publications, London (2006)
76. Karr, C.L., Yakushin, I., Nicolosi, K.: Solving inverse initial-value, boundary value problems via genetic algorithm. Eng. Appl. Artif. Intell. **13**, 625–633 (2000)
77. Kauati, A.T., Silva Neto, A.J., Roberty, N.C.: A source-detector methodology for the construction and solution of the one-dimensional inverse transport equation. Inverse Probl. Eng. **9**, 45–66 (2001)
78. Kennedy, J., Eberhart, R.C.: Swarm Inteligence. Morgan Kaufmann, New York (2001)
79. Kim, U., Kaluarachchi, J.J.: Application of parameter estimation and regionalization methodologies to ungauged basins of the Upper Blue Nile River Basin. Ethiopia. J. Hydrol. **362**, 39–56 (2008)

80. Kinoshita, S.K., Marques, P.M.A., Frère, A.F.: Use of principal component analysis and artificial neural networks for the classification of nodules in digitized mammograms. In: Anals of the XVII Brazilian Congress of Biomedical Engineering, Florianópolis, n. 1, pp. 1214–1216 (2000). (in Portuguese)
81. Kirkpartrick, S., Gelatt Jr., C.D., Vecchi, M.P.: Optimization by simulated annealing. Science **220**, 621–630 (1983)
82. Knupp, D.C., Silva Neto, A.J.: Simultaneous identification of the optical thickness and space-dependent albedo using Bayesian inference. 21st International Congress of Mechanical Engineering, ABCM, Natal, Brasil (2011)
83. Knupp, D.C., Silva Neto, A.J.: Solution of the inverse radiative transfer problem of simultaneous identification of the optical thickness and space dependent albedo using Bayesian inference. Comput. Model. Eng. Sci. **96**(5), 339–360 (2013)
84. Knupp, D.C.: Theoretical-Experimental Analysis of Heat Transfer and Nanocomposites via Integral and Infrared Thermographic Transformation. 2010. 174 f. Dissertação (Mechanical Engineering Mastership) | Universidade Federal do Rio de Janeiro, Rio de Janeiro (2010)
85. Knupp, D.C., Silva Neto, A.J., Sacco, W.F.: Radiative properties estimation with the particle collision algorithm based on a sensitivity analysis. High Temperatures—High Pressures **38**(2), 137–151 (2009)
86. Knupp, D.C., Sacco, W.F., Silva Neto, A.J.: Direct and inverse analysis of diffusive logistic population evolution with time delay and impulsive culling via integral transforms and hybrid optimization. Appl. Mathem. Comput. **250**, 105–120 (2015)
87. Knupp, D.C., Silva, L.G., Bevilacqua, L., Galeão, A.C.N.R., Silva neto, A.J.: Inverse Analysis of a New Anomalous Diffusion Model Employing Maximum Likelihood and Bayesian Estimation, Chapter 7, pp. 89–104, in Mathematical Modeling and Computational Intelligence in Engineering Applications, (Eds.), ISBN: 978-3-319-38868-7, Ed. Springer (2016)
88. Knupp, P., Salari, K.: Verification of Computer Codes in Computational Science and Engineering. Chapman & Hall/CRC, Boca Raton (2003)
89. Lazaro, L.L.B., Giatti, L.L., Macedo, L.S.V., Oliveira, J.A.P.: Water-Energy-Food Nexus and Climate Change in Cities, Sustainable Development Goals Series, pp. 327, ISBN: 978-3-031-05472-3, Ed. Springer (2022)
90. Le Roux, J.P.: An integrated law of the wall for hydrodynamically transitional flow over plane beds. Sedimentary Geol. **163**(3–4), 311–321 (2004)
91. Liang, D., Wang, X., Falconer, R.A., Bockelmann-Evans, B.N.: Solving the depth-integrated solute transport equation with a TVD-MacCormack scheme. Environ. Model. Softw. **25**, 1619–1629 (2010)
92. Lima, E.B.: Inverse problems applied to the identification of hydrodynamic parameters of a Macaé estuary model. 2012. 103 p. Thesis (Computational Modelling Doctor). Rio de Janeiro State University, Instituto Politécnico, Nova Friburgo (2012). (in Portuguese)
93. Lima, E.B., Rodrigues, P.P.G.W., Silva Neto, A.J., Irizar-Mesa, M.F., Llanes-Santiago, O., Lugon JR., J.: Parameter estimation in model of estuarine hydrodynamics based on genetic algorithms. 20th International Congress of Mechanical Engineering, ABCM, Gramado, Brasil (2009)
94. Lima, E.B., Rodrigues, P.P.G.W., Silva Neto, A.J., Lugon Jr., J., Irizar Mesa, M.F., Llanes Santiago, O.: Coupling MOHID with Optimization Algorithms: Perspectives on the Development of Automatic Calibration Tools, in Ocean Modelling for Coastal Management—Case Studies with MOHID, Mateus, M. e Neves, R. (Eds.), ISBN: 978-989-8481-24-5, Editora IST Press, pp. 117–130 (2013)
95. Lima, E.B., Rodrigues, P.P.G.W., Silva Neto, A.J., Mesa, M.I., Santiago, O.L., Lugon Jr., J.L.: Inverse problem for parameter estimation. In: Model of Estuarine Hydrodynamics. In: IX Congresso Internacional de Ingeniería Hidráulica Y V Seminário Internacional Del Uso Integral Del Água. Santiago, Cuba: [s.n.] (2009)
96. Lima, E.B., Silva Neto, A.J., Rodrigues, P.P.G.W.: Two-dimensional simulation of flow and transport in the estuary of the Macaé River. In: Computational Modeling Meeting. Volta Redonda: [s.n.] (2008). (in Portuguese).

97. Liu, X.: Analytical solutions for steady two-dimensional suspended sediment transport in channels with arbitrary advection velocity and eddy diffusivity distributions. J. Hydraulic Res. **54**(4), 389–298 (2016). Taylor & Francis
98. Lugon Jr, J., Kalas, F.A., Rodrigues, P.P.G.W., Jeveaux, J.L., Gallo Neto, H., Juliano, M.M., Silva Neto, A.J.: Lagrangian trajectory simulation of floating objects in the state of São Paulo coastal region. Defect and Diffusion Forum (DDF) **396**, 42–49 (2019)
99. Lugon Jr., J., Ferreira, M.I.P., Lima, E.B., Rodrigues, P.P.G.W., Silva Neto, A.J.: A sensitivity analysis for estuarine hydrodynamics regarding measurements of free surface level and longitudinal velocity. 20th International Congress of Mechanical Engineering, ABCM, Gramado, Brasil (2009)
100. Lugon Jr., J., Juliano, M.M., Kyriakides, I., Yamasaki, E.N., Rodrigues, P.P.G.W., Silva Neto, A.J.: Environmental hydrodynamic modelling applied to extreme events in Caribbean and Mediterranean countries. Desalination Water Treatment **194**, 315–323 (2020)
101. Lugon Jr., J., Kalas, F.A., Rodrigues, P.P.G.W., Juliano, M.M., Gallo Neto, H., Jeveaux, J.L., Silva Neto, A.J.: Application of MOHID platform to simulate the drift of cetaceans stranded in coastal regions. A Case Study in the Paulista Coastline, Revista CEREUS **11**(3), 178–189 (2019)
102. Lugon Jr., J., Paiva, P.M., Silva, J.A.F., Silva Neto, A.J.: Sensitivity study for a well blowout Lagrangian model at campos basin using MOHID platform. 24th International Congress of Mechanical Engineering, ABCM, Curitiba, Brasil (2017)
103. Lugon Jr., J., Rodrigues, P.P.G.W., Silva Neto, A.J.: Assessment of dispersion mechanisms in rivers by means of an inverse problem approach. Inverse Probl. Sci. Eng. **16**(8), 967–979 (2008)
104. Lugon Jr., J., Silva Neto, A.J., Santana, C.C.: A hybrid approach with artificial neural networks, Levenberg-Marquardt and simulated annealing methods for the solution of gas-liquid adsorption inverse problems. Inverse Probl. Sci. Eng. **17**(1), 85–96 (2009)
105. Luus, R., Jaakola, T.H.I.: Optimization by direct search and systematic reduction of the size of search region. AIChE J. **19**, 760–766 (1973)
106. Macro Region Environmental Intermunicipal Consortium 5—MRA 5. Report of constitution of the Macaé basin committee. Macaé (2002). (in Portuguese)
107. Maliska, C.R.: Fundamentals of Computational Fluid Dynamics, The Finite Volume Method, in Fluid Mechanics and Its Applications, Vol. 135, Cham, Springer (2023)
108. Mann, P.S.: Introductory Statistics. 5th ed. Wiley, USA (2004)
109. Martin, J.L., Mccutcheon: Hydrodynamics and transport for water quality modeling. New York: Lewis (1998)
110. Martinez Gonzáles, Y., Martinez Rodrigues, J.B., Silva Neto, A.J., Rodrigues, P.P.G.W.: Estimation of open channels hydraulic parameters with the stochastic particle collision algorithm. J. Brazilian Soc. Mech. Sci. Eng. **36**(1), 69–77 (2014)
111. Martins, F.A.B.C.: Three-dimensional Mathematical Modelling of a Coastal and Estuarine Flows Using a Generic Vertical Coordinate Approach. Thesis (Mechanical Engineering Doctor) University of Lisbon, Lisbon, 263 p. (1999). (in Portuguese)
112. Martins, F.: Three-dimensional Mathematical Modeling of Coastal and Estuarine Flows using a Generic Vertical Coordinate Approach. Ph.D., Thesis, Technical University of Lisbon (1999). (in Portuguese)
113. Martins, F., Leitão, P., Silva, A., Neves, R.: 3D modeling in the Sado estuary using a new generic vertical discretization approach. Oceanol. Acta **24**, 51–62 (2001)
114. Miranda, L.B., de Castro, B.M., Kjerfve, B.: Principles of Physical Oceanography of Estuaries. Editora Universidade de São Paulo, São Paulo (2002). (in Portuguese)
115. Miranda, R., Braunschweig, F., Leitão, P., Neves, R., Martins, F., Santos, A.M.: A Costal integrated object oriented model. Hydraulic Engineering Software VIII, WIT Press. [S.l.]: [s.n.] (2000)
116. Montero, P.: Study of the hydrodynamics of the Ría de Vigo by means of a finite volumes model. Ph.D, Thesis, Universidad de Santiago de Compostela (1999). (in Spanish).

117. Montero, P., Gómez-Gesteira, M., Taboada, J.J., Ruiz-Villarreal, M., Santos, A.P., Neves, R.J.J., Prego, R., Pérez-Villar, V.: On residual circulation of Vigo Ría using a 3D baroclinic model. Boletín Instituto Español de Oceanografía, v. 15. SUPLEMENTO-1 (1999)
118. Moura Neto, F.D., Silva Neto, A.J.: An Introduction to Inverse Problems with Applications. Springer, Heidlberg (2013)
119. Munavalli, G.R., Kumar, M.S.M.: Water quality parameter estimation in a distribution system under dynamic state. Water Res. **39**(18), 4287–4298 (2005)
120. Naveira-Cotta, C.P., Orlande, H.R.B., Cotta, R.M.: Integral transforms and Bayesian inference in the identification of variable thermal conductivity in two-phase dispersed systems. Numer. Heat Transfer Part B: Fund. Taylor & Francis **57**(3), 173–202 (2010)
121. Oliveira, M., Souza, N., Lugon Jr., J., Silva Neto, A.J., Neves, R.: Mathematical modelling and simulations for microplastic environmental research: a systematic review. Environ. Rev. **33**, 1–18 (2025)
122. Orlande, H.R.B., Colaço, M.J., Cotta, C.P.N., Guimarães, G., Borges, V.L.: Inverse Problems in Heat Transfer. São Carlos, SP: SBMAC, 2011. v. 51. (Notes in Applied Mathematics, v. 51). (in Portuguese).
123. Paiva, P.M., Lugon Jr., J., Barreto, A.N., Silva, J.A.F., Silva Neto, A.J.: Comparing 3D and 2D computational modeling of an oil well blowout using MOHID platform—a case study in the campos basin. Sci. Total Environ. **595**, 633–641 (2017)
124. Parolin, R.S., Marón Dominguez, D.E., Llanes Santiago, O., Silva Neto, A.J., Rodrigues, P.P.G.W.: Sensitivity analysis and estimation of a source of contaminant in the estuary of the Macaé River. RBRH—Brazilian Water Resour. J. **20**(1), 24–33 (2015). (in Portuguese)
125. Parolin, R.S., Silva Neto, A.J., Rodrigues, P.P.G.W.: Determination of a Pollutants Source in an Estuary through the Inverse Problem Technique, 22nd International Congress of Mechanical Engineering, ABCM, Ribeirão Preto, Brasil, pp. 10597–10604 (2013)
126. Parolin, R.S., Silva Neto, A.J., Rodrigues, P.P.G.W., Llanes Santiago, O.: Estimation of a contaminant source in an estuary with an inverse problem approach. Appl. Mathem. Comput. **260**, 331–341 (2015)
127. Patankar, S.V.: Numerical Heat Transfer and Fluid Flow. Hemisphere, New York (1980)
128. Pérez-Villar, V.: Integrated Management of the Galician Maritime Terrestrial Space: Numerical Modeling. Final report by the Grupo de Física Non Lineal, Consellería de Pesca, Marisqueo e Acuicultura. Xunta de Galicia (1999). (in Spanish).
129. Pinto, L., Fortunato, A.B., Zhang, Y., Oliveira, A., Sancho, F.E.P.: Development and validation of a three-dimensional morphodynamic modelling system for non-cohesive sediments. Ocean Model. Elsevier **57**, 1–14 (2012)
130. Rao, S.S.: Engineering Optimization, Theory and Practice. Third Edition. ed. [s.l.]: John Wiley & Sons, Inc. (1996)
131. Rennó, C.D., Soares, J.D.: Hydrological models for environmental management. São José dos Campos: Instituto Nacional de Pesquisas Espaciais (2000). 60 p. (in Portuguese)
132. Rijn, L.C.V.: Mathematical modeling of suspended sediment in nonuniform flows. J. Hydraulic Eng. Amer. Soc. Civil Eng. **112**(6), 433–455 (1986)
133. Rodrigues, P.P.G.W., Lugon Jr, J., Tavares, M.C.: Importance of Mathematical Models in Environmental Studies—Case Study: the Rio Macaé. Boletim do Observatório Ambiental Alberto Ribeiro Lamego, v. 1, n. 2, p. 115–128 (2007). (in Portuguese).
134. Rodrigues, P.P.G.W., Telles, W.R., Parolin, R.S., Lugon Jr., J., Silva Neto, A.J.: Contaminant Transport Modeling and Source Identification with a Case Study in the Lower Macaé, Chapter, pp. 435–454, in Engineering and Environmental Sciences: Contributions to Ecosystem Management, (Ed.), ISBN: 978-85-99968-58-1, Editora Essentia (2019). (in Portuguese)
135. Rodrigues, P.P.W.G., Lima, E.B., Silva Neto, A.J., Lugon Jr, J.: Modelling and sensitivity analysis for estuarine hydrodynamics. In: IX SIMPOSIO INTERNACIONAL DE AUTOMATIZACIÓN. Havana, Cuba (2009)
136. Rodrigues, P.P.G.W., Barnes, J., Upstill-Goddard, R.C.: Simulating estuarine nitrous oxide production by means of a dynamic model. Mar. Pollut. Bull. **54**, 164–172 (2007)

137. Rosman, P.C.: Circulation models in shallow bodies. In: Vieira da Silva, R.C. (Ed.). Rio de Janeiro: Numerical Methods in Water Resources, ABRH (1989). (in Portuguese).
138. Runkel, R.L.: Solution of the advection-dispersion equation: continuous load of finite duration. J. Environ. Eng. Amer. Soc. Civil Eng. **122**(9), 830–832 (1996)
139. Sacco, W.F., Oliveira, C.R.E., Pereira, C.M.N.: Two stochastic optimization algorithms applied to nuclear reactor core design. Prog. Nucl. Energy **48**(6), 525–539 (2006)
140. Sacco, W.F., Oliveira, C.R.E.: A new stochastic optimization algorithm based on particle collisions. Transactions of the America Nuclear Society, San Diego (2005)
141. Sanches, E.L., Knupp, D.C., Abreu, L.A.S., Mascout, F.S., Silva Neto, A.J.: Reconstruction of Spacewise varying boundary heat flux employing a combination of explicit estimation with Markov chain Monte Carlo method. 25th International Congress of Mechanical Engineering, ABCM, Uberlândia, Brasil (2019)
142. Santos, I., Fill, H.D., Sugai, M.R.B., Buba, H., Kishi, R.T., Marone, E., Lautert, L.F.C.: Applied Hydrography. Curitiba: Instituto de Tecnologia para o Desenvolvimento (2001). (in Portuguese)
143. Schnoor, J.L.: Environmental Modeling—Fate and Transport of Pollutants in Water, Air and Soil. Wiley-Interscience (1996)
144. Schwaab, M., Pinto, J.C.: Analysis of Experimental Data: I. Fundamentals of Statistics and Estimation of Parameters. Rio de Janeiro: Editora E-papers (2007). (in Portuguese)
145. Shanahan, P., Gaudet, S.C.: Mixing and Transport of Pollutants in Surface Water, Chapter 1, Section 5B, in Standard Handbook of Environmental Science, Health and Technology, Jay H. Lehr, editor. 2000. McGraw Hill, New York (2000)
146. Shen, J.: Optimal estimation of parameters for an estuarine eutrophication model. Ecol. Model. **191**(3–4), 521–537 (2006)
147. Shen, J., Jiab, J., Sissona, G.M.: Inverse estimation of nonpoint sources of fecal coliform for establishing allowable load for Wye River. Maryland. Water Res. **40**, 3333–3342 (2006)
148. Silva Neto, A.J.: Explicit and implicit formulations for the inverse radiative transfer problems. Proceedings of the 5th World Congress on Computational Mechanics, Vienna, Austria (2002)
149. Silva Neto, A.J., Soeiro, F.J.C.P.: Inverse problem of space dependent albedo estimation using hybrid optimization methods. Proceedings of the 6th World Congress on Structural and Multidisciplinary Optimization, Rio de Janeiro (2005)
150. Silva Neto, A.J., Becceneri, J.C., Campos Velho, H.F.: Computational Intelligence Applied to Inverse Problems in Radiative Transfer. Springer, Cham (2023)
151. Silva Neto, A.J., McCormick, N.J.: An explicit formulation based on the moments of the exit radiation intensity for the one-dimensional inverse radiative transfer problem. Proceedings of the 4th International Conference on Inverse Problems in Engineering: Theory and Practice, Angra dos Reis, Brazil, Vol. II, pp. 347–354 (2002)
152. Silva Neto, A. J.; Roberty, N. C.; Pinheiro, R. P. F.; Alvarez Acevedo, N.I.: Inverse problems explicit and implicit formulations with applications in engineering, biophysics and biotechnology. Inverse Probl. Sci. Eng. **15**(4), 373–411 (2007)
153. Silva Neto, A.J., Soeiro, F.J.C.P.: Estimation of phase function of anisotropic Scatering with a combination of gradient based and stochastic global optimization methods. Proceedings of the 5th World Congress on Computational Mechanics, Vienna, Austria (2002)
154. Silva Neto, A.J., Soeiro, F.J.C.P.: Solution of implicitly formulated inverse heat transfer problems with hybrid methods. Comput. Fluid Solid Mech., 2369–2372. Elsevier, Cambridge (2003)
155. Silva, J.T.: Determination of River Flow Rates Using the Chemical Method: Application to the Upper Macaé (2009). 52 p. Monography (Mechanical Engineering Graduation). Rio de Janeiro State University, Instituto Politécnico, Nova Friburgo (2009). (in Portuguese)
156. Silva, W.B., Dutra, J.C.S., Abreu, L.A.S., Knupp, D.C., Silva Neto, A.J.: Estimation of Timewise varying boundary heat flux via Bayesian filters and Markov Chain Monte Carlo Method, Chapter 8, pp. 137–153, in Llanes Santiago, O., Cruz Corona, C., Silva Neto, A.J., Verdegay, J.L. (Eds.): Computational Intelligence in Emerging Technologies for Engineering Applications. Studies in Computational Intelligence, Vol. 872, ISBN: ISBN: 978-3-030-34408-5 (Hardcover), 978-3-030-34409-2 (eBook), Ed. Springer, Cham (2020)

157. Smith, G.F.: Numerical Solution of Partial Differential Equations: Finite Difference Methods, 3rd ed. Oxford University Press, New York (2004)
158. Sodré, U.: Mathematical models. Londrina: Universidade Estadual de Londrina (2007). (in Portuguese)
159. Soeiro, F.J.C.P., Stutz, L.T., Tenenbaum, R.A., silva neto, A.J.: Stochastic and hybrid methods for the identification in the Bouc-Wen model for the magneto-rheological dampers. J. Phys. Conf. Series **135**, 012093 (2008)
160. Souza, A.L.R.: Study of the Incipient Movement of Non-Cohesive Sediments in Flows with Free Surface. M.S. dissertation—Rio de Janeiro: UFRJ/COPPE (2010). (in Portuguese)
161. Souza, A.L.R., Knupp, D.C., Rodrigues, P.P.G.W., Silva Neto, A.J.: Integral transforms and inverse problems applied to parameter estimation in a 2D model of non-cohesive sediment transport. Proceeding Series of the Brazilian Society of Computational and Applied Mathematics (2017)
162. Souza, A.L.R., Silva Neto, A.J., Knupp, D.C., Rodrigues, P.P.G.W.: Parameter estimation in a simulation of two-dimensional sediment transport using integral transform and Bayesian inference. 25th International Congress of Mechanical Engineering, ABCM, Uberlândia, Brasil (2019)
163. Souza, A.L.R.: Generalized Integral Transformation and Bayesian Inference Applied to Inverse Problems in Non-Cohesive Sediment Transport. Ph.D. thesis, Nova Friburgo: IPRJ/UERJ (2018). (in Portuguese)
164. Souza, E.P.D.: Evaluation of dispersive mechanisms in rivers through inverse problems (2009). 95 p. Master Dissertation (Computational Modelling Graduate Program). Rio de Janeiro State University, Instituto Politécnico, Nova Friburgo (2009). (in Portuguese)
165. Souza, N., Lugon Jr., J., Fernandes, A.M., Neves, R.J.J., Silva Neto, A.J.: The Use of general inverse problem platform (GRIPP) as a robust backtracking solution. Coasts **5**(6), 50100006.1–19 (2025)
166. Souza, N., Lugon JR., J., Fernandes, A.M., Neves, R.J.J., Silva Neto, A.J.: Inverse Problem Solution for Microplastic Emission Source Area Estimation in MOHID: The Sepetiba Bay Case Study, Ciência e Natura, Vol. XX, pp. eYY-YY (2025)
167. Souza, N., Lugon JR., J., Yamasaki, E., Kyriakides, I., Silva Neto, A.J.: An Assessment of relative potential impacts to Cyprus' shoreline due to oil spills in the eastern Mediterranean sea. J. Integrated Coastal Zone Manag. **23**(1), 27–40 (2023)
168. Stephany, S., Becceneri, J.C., Souto, R.P., Campos Velho, H.F., Silva Neto, A.J.; A pre-regularization scheme for the reconstruction of a spatial dependent scattering albedo using a hybrid ant colony optimization implementation. Appl. Mathem. Model. **34**, 561–572 (2010)
169. Steyaert, L.T.A.: Perspective on the state of environmental simulation modelling. In: Goodchild, M.F., Parks, B.O., Steyaert, L.T. (Eds.) Environmental modeling with GIS. New York, Oxford University Press. Cap. 3, pp. 16–30 (1993)
170. Storn, R., Price, K.: Differential evolution: a simple and efficient adaptive scheme for global optimization over continuous spaces. J. Global Optim. **12**, 1–16 (1997)
171. Streeter, V.F., Wylie, E.B., Bedford, K.W.: Fluid Mechanics, McGraw-Hill International Editions. 9th ed. Singapore. 740p. (1998)
172. Strub, I.S., Percelay, J., Stacey, M.T., Bayen, A.M.: Inverse estimation of open boundary conditions in tidal channels. Ocean Model. **29**(1), 85–93 (2009)
173. Taboada, J.J.: Application of numerical models for the study of hydro-dynamics and particle fluxes in the Mediterranean Sea. Ph.D. 103 Thesis, Universidad de Santiago de Compostela (1999). (in Spanish)
174. Taboada, J.J., Prego, R., Ruiz-Vellarreal, M., Montero, P., Gómezgesteira, M., Santos, A., Pérez-Villar, V.: Evaluation of the seasonal variations in the residual patterns in the Ría de Vigo (NW Spain) by means of a 3D baroclinic model. Estuar. Coast. Shelf Sci. **47**, 661–670 (1998)
175. Tannehill, J.C., Anderson, D.A., Pletcher, R.H.: Computational Fluid Mechanics and Heat Transfer, 2nd edn. Taylor & Francis, Washington (1997)

176. Telles, W.R.: Simulation of the two-dimensional horizontal transport of conservative substance (2009). 88 p. Dissertation (Computational Modelling Mastership). Rio de Janeiro State University, Instituto Politécnico, Nova Friburgo (2009). (in Portuguese).
177. Telles, W.R., Brandão, D.N., Lugon Jr., J., Rodrigues, P.P.G.W., Silva Neto, A.J.: Computational modelling and simulation of surface Waterflood in mountainous urban watersheds with the MOHID platform: case study Nova Friburgo, Brazil, Chapter, pp. 125–143, in Towards Mathematics, Computers and Environment: A Disasters Perspective, (Eds.), ISBN: 978-3-030-21204-9, Editora Springer (2019)
178. Telles, W.R., Rodrigues, P.P.G.W., Silva Neto, A.J.: Automatic Calibration of the MOHID Platform Employing a Stochastic Optimization Method and Real Data from an Extreme Weather Event in Nova Friburgo-RJ: Part 1—Preparation of the Digital Terrain Model and Acquisition of Experimental Precipitation and Level Data, Revista Internacional de Métodos Numéricos para Cálculo y Diseño en Ingeniería, v. 33, n. 3–4, pp. 164–170 (2017). (in Portuguese).
179. Telles, W.R., Rodrigues, P.P.G.W., Silva Neto, A.J.: Automatic Calibration of the MOHID Platform Employing a Stochastic Optimization Method and Real Data from an Extreme Weather Event in Nova Friburgo-RJ: Part 2—Sensitivity Analysis and Estimation of Hydrological Parameters, Revista Internacional de Métodos Numéricos para Cálculo y Diseño en Ingeniería, v. 33, n. 3–4, pp. 204–211 (2017). (in Portuguese)
180. Telles, W.R., Rodrigues, P.P.G.W., Silva Neto, A.J.: Automatic calibration of a simulator applied to a mountain river using experimental precipitation and level data—case study: Córrego D'Antas, RJ. RBRH—Brazilian Water Resour. J. **21** (1), 143–151 (2016). (in Portuguese)
181. Telles, W.R., Rodrigues, P.P.G.W., Silva Neto, A.J.: Use of Geographic Information Systems in the Delimitation of Watersheds, Boletim do Observatório Ambiental Alberto Ribeiro Lamego **7**(1), 19–30 (2013). (in Portuguese).
182. Telles, W.R., Rodrigues, P.P.G.W., Silva Neto, A.J., Llanes Santiago, O., Irizar Mesa, M., Lugon Jr., J.: Simulation of a pollutant plume in the Macaé River using artificial neural networks. RBRH—Brazilian Water Resour. J. **18**(2), 165–174 (2013). (in Portuguese)
183. Telles, W.R., Silva Neto, A.J., Rodrigues, P.P.G.W.: Modelling the flow of a river using the MOHID platform. 22nd International Congress of Mechanical Engineering, ABCM, Ribeirão Preto, Brasil, pp. 8539–8547 (2013)
184. Tenenbaum, R.A., Fernandes, K.M., Stutz, L.T., Silva Neto, A.J.: Damage identification in bars with a wave propagation approach and a hybrid optimization method. Shock Vib. **19**, 301–321 (2012)
185. Tikhonov, A.N., Arsenin, E.: Solutions of III-Posed Problems.V. H. Winston, Washington (1977)
186. Tonsig, S.L.: Multilayer Artificial Neural Networks and the Backpropagation Algorithm. São Paulo: Universidade de São Paulo (2000) (in Portuguese)
187. Tsanis, I., Shen, H., Valeo, C.: Environmental Hydraulics: Hydrodynamic and Pollutant Transport Models of Lakes and Coastal Waters with Engineering Applications, vol. 56. Elsevier, Amsterdam (2007)
188. Tucci, C.E.M.: Hydrology: Science and Application. Rio Grande do Sul: Editora da UFRGS, 2007. (ABRH Water Resources Collection). ISBN: 978-85-7025-924-0. (in Portuguese)
189. United Nations, The United Nations World Water Development Report 2024: Water for Prosperity and Peace. UNESCO, Paris (2024). ISBN: 978-92-3-100657-9
190. Vallino, J.J., Hopkinson, C.S.: Estimation of dispersion and characteristic mixing times in plum island sound estuary. Estuar. Coast. Shelf Sci. **46**, 333–350 (1998)
191. Versteeg, H.K., Malalasekera, W.: An introduction to Computational Fluid Dynamics: The Finite Volume Method. 2nd Edition, Pearson Education Limited (2007)
192. Von Sperling, M.: Studies and modeling of river water quality. 2o. ed. Belo Horizonte: Editora UFMG, (2014). v. 7. (Principles of Biological Treatment of Wastewater, v. 7). (in Portuguese)
193. Wolfram Research, Inc., Mathematica, Version 13.2, Champaign, IL (2022)
194. Wrobel, L.C. Introduction to numerical methods. In R.C.V. da Silva, R.C.V. (Ed.), Numerical Methods in Water Resources. ABRH, Porto Alegre (1989). (in Portuguese).

195. Yamamoto, Y., Nikiforuk, P.: A new supervised learning algorithm for multilayered and interconnected neural networks. IEEE Trans. Neural Networks **11**(1), (2000)
196. Yang, Z., Hamrick, J.M.: Optimal control of salinity boundary condition in a tidal model using a variational inverse method. Estuar. Coast. Shelf Sci. **62**(1–2), 13–24 (2005)
197. Yotsukura, N., Cobb, E.D.: Transverse diffusion of solutes in natural streams. Dispersion in Surface Water. U.S. Geological Survey Professional Paper 582-C (1972)
198. Zienkiewicz, O.C., Taylor, R.L.: The Finite Element Method. Vol. 3: Fluid Dynamics, 5th ed. Butterworth-Heinemann, Oxford (2000)

Index

J. Lugon Junior et al., *Inverse Problems in the Environmental Modelling of Rivers and Estuaries*, Studies in Systems, Decision and Control 246,
https://doi.org/10.1007/978-3-032-09937-2

MIX
Papier aus verantwortungsvollen Quellen
Paper from responsible sources
FSC® C105338

If you have any concerns about our products,
you can contact us on
ProductSafety@springernature.com

In case Publisher is established outside the EU,
the EU authorized representative is:
Springer Nature Customer Service Center GmbH
Europaplatz 3, 69115 Heidelberg, Germany

Printed by Libri Plureos GmbH
in Hamburg, Germany